# Methods to Assess Quality and Stability of Oils and Fat-Containing Foods

# Methods to Assess Quality and Stability of Oils and Fat-Containing Foods

Editors

**Kathleen Warner**
USDA, ARS
Peoria, Illinois

**N.A. Michael Eskin**
University of Manitoba
Winnipeg, Manitoba, Canada

AOCS PRESS
Champaign, Illinois

The paper used in this book is acid-free and falls within the guidelines established to ensure permanence and durability.

Methods to assess quality and stability of oils and fat-containing
foods/editors, Kathleen Warner, N.A. Michael Eskin.
p. cm.
Includes bibliographical references and index.
ISBN 0-935315-58-6 (alk. paper)
1. Oils and fats, Edible—Quality control. 2. Food adulteration and inspection. 3. Food industry and trade—Quality control.
I. Warner, Kathleen. II. Eskin, N. A. M. (Neason Akivah Michael)
TX560.03M48 1994
664′.3—dc20 94-48192
CIP

Printed in the United States of America with vegetable oil–based inks.

00 99 98 97 96 95 5 4 3 2 1

*This book is dedicated to*

*Dr. Sybil James*

*Reader in Biochemistry (Retired)*

*University of Birmingham, England*

*and*

*Dr. Edwin N. Frankel*

*Department of Food Science and Technology*

*University of California, Davis*

# Preface

This monograph is based on a successful AOCS short course held in Chicago, Illinois, prior to the 1991 Annual Meeting. Interest is particularly high for short courses dealing with valid and reliable methods to assess oxidation of fats and oils. This monograph was written to meet the needs of the growing number of laboratory scientists in quality control, product development, and research for industry, government and academia who are establishing protocols to determine oil quality and stability. The information presented here should assist in selecting methods that best represent the true state of quality and stability in oxidized lipids.

The first two chapters provide the reader with a broad perspective from which to consider the information in the subsequent methodology chapters. Chapter 1 presents a historical review of the development of methods to monitor fats and oils form the 1940s through the 1970s. All readers, but especially novices in lipid oxidation, will appreciate the origins of this methodology and the advances in methods to assess oxidation. Chapter 2 reviews factors affecting oil quality and stability, based on the literature and the author's own experiences.

In subsequent chapters, scientists who are authorities in measuring lipid oxidation describe the primary methods, as well as the advantages and limitations of these procedures. Not every method is included, only those that are most representative of oxidative deterioration in foods. Although chemical analyses are the foundation of many quality control laboratories, sensory analysis is the ultimate analytical test of oil quality for food-grade products. All aspects of sensory analysis are included, such as selecting and training panelists, developing sensory panel facilities, preparing testing samples, and designing and choosing testing methods. We have included those instrumental and chemical tests that relate most closely to actual sensory quality and stability, such as gas chromatographic analysis of volatile compounds, peroxide value, conjugated dienes, carbonyl value, oxygen uptake, and anisidine value. The best alternatives to sensory analysis, the three types of gas chromatographic analysis of volatile compounds are presented, including direct injection, static headspace and dynamic headspace. We highly recommend using more than one method and suggest a combination of at least two or three as the preferred protocol.

Finally, no book on methods would be complete without the information presented in Chapters 10 and 11 on the critical procedures essential before and after any analysis, including developing an experimental design, oxidizing samples under proper conditions, and statistically analyzing the data.

*Kathleen Warner*
*N.A. Michael Eskin*

# Contents

# Methods to Assess Quality and Stability of Oils and Fat-Containing Foods

Chapter 1

# Historical Glimpses of Analytical and Quality Assessment Methods for Fats and Oils

**Herbert J. Dutton**

The Hormel Institute, University of Minnesota, Austin, Minnesota 55912, USA.

*"Jeder Wissenschaftliche Fortschritt ist ein Fortschritt der Methode."* (1)

This review presents seven glimpses of progress rather than a continuum of history. The account, admittedly subjective, projects the author's viewpoint and experience beginning in an industrial laboratory in 1936 and continues with developments in fat analysis and sensory evaluation over a period of nearly 60 years.

## An Industrial Food Lab Out of the 1930s

The analytical methodology of the 1930s is perhaps best illustrated by the photograph in Figure 1.1. Above the wood surfaced laboratory desk can be seen the shelf with 5-gallon glass carboys with standardized solutions of acid, alkali, thiosulfate, and other chemicals, each connected to a burette for determining acid value (2), free fatty acid (3), ammonia nitrogen (4), and peroxide value (5). In addition to controlling boiler water (hardness, oxygen content, phosphate, and tannin), a pioneering sewage pretreatment plant was monitored with a variety of tests including biological oxygen demand (BOD).

The lower bank of flasks on the left was for digesting foods and feedstuff protein in sulfuric acid. After neutralization, the upper bank was used for ammonia distillation, required for the Kjeldahl protein N determination. On the opposite sides of the laboratory (not shown in Figure 1.1) were the extractors for determining fat content, the vacuum oven for determining moisture, the "Swift stability test" bath, and the analytical balance. A set of calibrated weights from the National Bureau of Standards were used, and tenths of milligrams were calculated by the all-but-forgotten method of swings. On the floor above this laboratory was housed the administrative offices of the company. Here, taste testing was conducted on occasion and is discussed later.

It was the author's dubious distinction to have nearly burned down this wooden structure. When cleaning the desktop of grease and oil after the day's work, a petroleum ether-wetted cloth was routinely used to remove any spills from the black bench top. One afternoon, I was alone in the lab and had almost completed the operation, when at the end of the bench behind the place where Dan H. Nelson on the right in Figure 1.1 is standing, I saw to my horror a lighted Bunsen burner. Before I could act, the fumes ignited and yellow smoky flames licked the wood ceiling; the

**Figure 1.1.** A meat packer's laboratory in the 1930s.

fire died down as rapidly as it rose, leaving a settling cloud of carbon particles. I had extinguished the ether-soaked cloth in the corner sink and was cooling my singed hands under the faucet when Nelson entered and observed "Kinda dusty in here today." I laconically agreed.

Some explanation is needed to understand the peculiar role of science in the meat-packing industry in the 1930s. Dr. Nelson and I, a student and part time pair of hands at the B.A. degree level, were the only two technically trained people in this organization. Curing hams, for example, was done by "secret" formulas (which, of course, every company in the industry knew). To Nelson, came the responsibility of staking his scientific reputation on his recommendations, such as that the less expensive beet sugar could be exchanged for cane sugar in the ham cures. He had to explain to the administration why meat-storage coolers had mold-inducing water on the floor. Our allegedly toxic bacon brought in by the city chemist was tested simply by frying it in an old iron skillet. Both Nelson and the city chemist would then eat it, knowing that the frying temperature would destroy any biotoxins present.

The role of the chemist was regarded suspiciously by the department foremen. Confided to me by the Russian-born foreman of curing operations concerning our analysis of ham-curing pickle, he said "Doc, He no have to find 'em; he know what he put 'em; he just book 'em up-down; show 'em B___ (the boss)." Translated, this meant that Nelson doesn't have to find salt, sugar, nitrate (and nitrite) in the pickle; he knows what he added; he just writes it in a record book and shows it to the plant manager. Despite this distrust, the lab constituted a neutral meeting ground where warring government inspectors, department foremen, and administrators could communicate with fewer inhibitions.

The Taste Testing, as it was called, was conducted in the administration offices. Hams from various experimental cures were cut for tasting by management personnel. In these roundtable discussions, I learned that Nelson could never get the business manager to do a "blind" test. His judgment, as well as his sensory responses, were based on how long the ham had been in cure (i.e., the cost of the cure).

You are probably thinking I have spent an inordinate amount of space on anecdotal material, but I have done so to lay a basis in the past from which to view the present state of analytical methodology and quality assessment. The beginnings of current methods for analysis were then present. The American Meat Institute was evaluating the Swift Stability Test for lards—the precursor of the current fat stability, Active Oxygen Method (AOM) test (6). We confirmed that the time of induction for peroxide development was a helpful index of stability in lard products. Because the peroxide value (PV) in lard rose rapidly once started (autocatalysis), almost any arbitrary PV could be used, 100, 200, 500 PV for the end point, and arrive at essentially the same time of lard stability. One could also smell the rancidity at the exit tube as a "quick and dirty" monitor.

At this period of time, we had a refractometer to measure the refractive index of oil (7). I am not totally sure why we had a refractometer in an oil lab. Color tubes and standards were available with which to compare tallow color. Colorimetry had yet to be accepted, and spectrophotometers had yet to be invented.

The solutions for the Hanus or Wijs iodine value (8) were available, but in our industry we had little need for fatty acid compositional information. Years later, I rescued two Wijs iodine value flasks with their unique form from the discard pile, because by then I recognized them as symbolic of that period. It is ironic that today when an iodine value (IV) is required, it is most probably calculated from gas chromatographic data. However, IV was our grandfather's gas chromatographic-mass (GC–MS) spectrophotometric analysis, and this brings me to the next series of glimpses, the 1950s.

## Alkali Conjugation — Spectrophotometry

From IV one could speculate whether the fatty acids of an unknown oil were monosaturated (oleic acid 89.87 IV), diunsaturated (linoleic acid 181.69 IV), or triunsaturated (linolenic acid 273.51 IV) or guess whether the natural mixture was complicated by the presence of saturated acids. This uncertain picture was clarified by the application of the newly invented absorption spectrophotometer; Beckman's Model DU recently celebrated its 50th anniversary.

By measuring spectral absorption at 232 mμ before and after alkali conjugation, one could assess how much conjugated diene, linoleic acid, was present. If conjugatable trienes were present, a measurement at 268 mμ was also made. The linolenic acid was calculated, and an appropriate correction in the conjugatable diene made. This was a notable achievement (9) for lipid analysis because now a rapid independent analysis could be given to a variety of unsaturated fatty acids occurring in animal and vegetable oils.

At this point, the development was yet incomplete because the monoenoic and saturated acids still were not accounted for. The ingenuous solution to this dilemma lay in considering both IV and spectrophotometric data together and calculating oleic acid. Thus, by the combined iodine value–alkali conjugation–spectrophotometric methods oleic, linoleic, and linolenic acids were determined. The difference between the sum of unsaturated acids and 100% was the saturated acid content. What an achievement in methodology and how important to the growing compositional studies of lipids of that day.

## Differential Migration Processes

Enter now a new, diverse, ubiquitous, and multifaceted methodology in lipid analysis, with ancient origins but with popularity and utility that suddenly increased in the lipid analytical scene. Included under this heading are the now common words and acronyms of the chemists' lexicon: paper chromatography (PC), adsorption analysis (AA), thin-layer chromatography (TLC), countercurrent distribution (CCD), counter double counter distribution (CDCD), liquid chromatography (LC), high performance liquid chromatography (HPLC), gas chromatography (GC), and capillary gas chromatography (CGC) to name a few.

Paper chromatography, the technique first described by Tswett, was primarily the separation of pigments as the *chrom* or color prefix denotes. Apparatus of the early 1940s for column chromatography and the determination of carotene in dehydrated vegetables is shown in the collage (Figure 1.2a). A paper of mine in 1944, originally titled "Chromatography of Colorless Compounds" was changed to read "Adsorption Analysis of Colorless Lipids" to avoid the obvious oxymoron of writing about colorless color (10). The subtitle "Resolution of Stearic and Oleic Acid" of the paper described an early chromatographic separation of these two fatty acids. Equally important, as shown in the collage (Figure 1.2b), it had the basic elements of HPLC, for example, pressure, solvent, column, and flow through differential refractometric monitoring (sensitivity $2 \times 10^{-6}$) (11).

Thin-layer chromatography was described in the United States as early as 1950 by Kirschner (12), and his version carried a host of monachers such as "chromato-strips," and "chromato-bars" developed in "chromato-cabs." Thin-layer chromatography obtained wide recognition only after a facile procedure and useful equipment described by Stahl became available to make chromatographic plates (13). In the hands of Mangold (14) and others at the Hormel Institute, reversed phase, silver ion (15), and two-dimensional variations, among others, were applied to lipids (Figure 2c). After all these years, TLC remains a procedure of choice in many laboratories for the separation of lipid classes, geometric and positional isomer resolution, and systems of isomers isologous in number of double bonds.

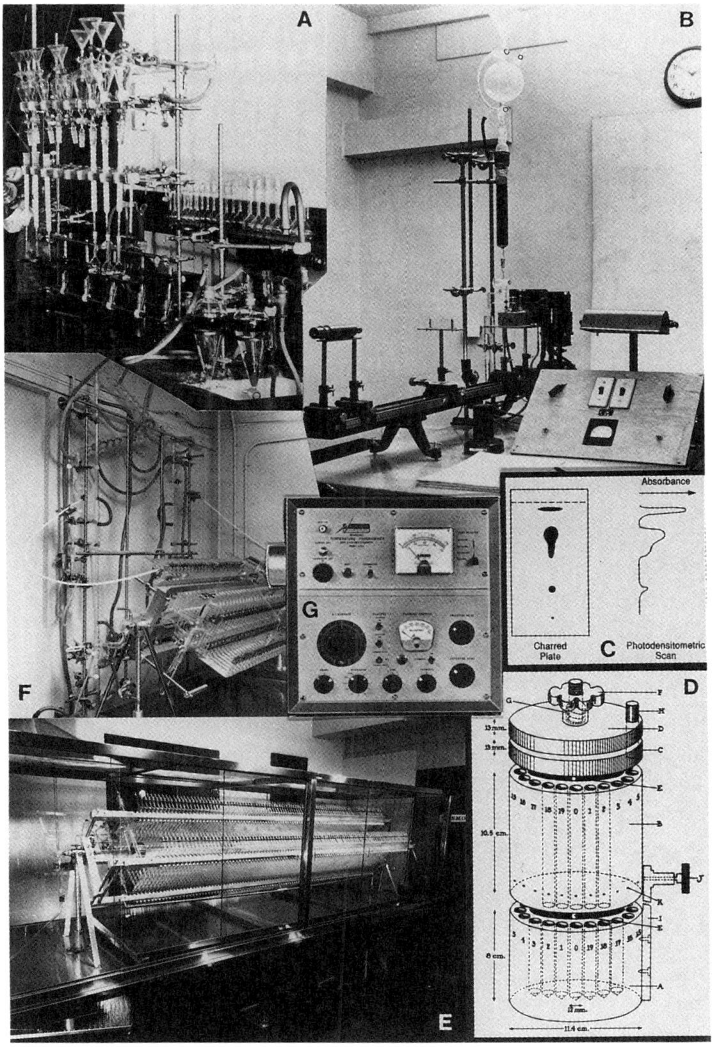


**Figure 1.2.** Differential migration processes. (a) Apparatus for the analysis of carotene in fresh and dehydrated vegetables by adsorption column chromatography. (b) Forerunner of HPLC with pressure, solvent, column, and differential refractometer to separate fatty acids. (c) The ubiquitous TLC methodology. (d) Original metal version of countercurrent distribution (CD). (e) Automatic 200 tube CD. (f) Counter double current distribution (CDCD) with continuous solvent and product recovery. (g) An early "Aerograph" gas chromatograph with thermoconductivity detector and 6-volt storage battery power supply.

Another unit of the collage of differential migration processes is that based on liquid/liquid extraction. Apparatus invented by Post and Craig (17) first in metal (Figure 1.2d) and then in glass (Figure 1.2e), countercurrent distribution was first used with fatty acids by their colleague, Ahrends (18) at the Rockefeller Institute. When the "glass pipe organ" applied to the separation of triglycerides in a 200 glass tube model, remarkable separations were obtained, upsetting Hildich's widely accepted theories on "even" and "random" distribution of fatty acids within triglyceride molecules (1920).

In another section of the collage (Figure 1.2f) is shown the CDCD where the two immiscible solvents moved stepwise and countercurrently after the shaking and settling stages. In a closed system of solvent steam distillation and product recovery (21), 500 g of pure methyl linolenate were recovered from linseed methyl esters in the first working week of its operation. At the price of the pure compound, the purchase price of the CDCD equipment was nearly returned in the first week. Indications were that fatty acids more unsaturated than linolenic presented even more favorable separation systems. Although generally replaced by the more popular "preparative" HPLC procedures, I contend that it is still a choice preparation method. One problem would be that with such a high production rate, supply for high-purity fatty acids for research purposes would soon exceed demand.

The next significant picture of the collage dealing with differential migration processes is GC (Figure 1.2g). In the 1950s, one of the missions undertaken by the informal Ad Hoc Committee on Gas Chromatography was to try to tell the GC equipment manufacturers that beyond the currently successful applications to petroleum-product separation was the potentially large market of application to medical, biological, and lipid research—if only they would raise the high-temperature limit on the thermoconductivity detectors with improved electrical insulators. Out of this group came the famed selective polyester liquid phases, such as Reoplex 400, LHC-2-R-446, and ethylene glycol succinate, (22), and from S. Lipsky (23) the first 500 ft stainless steel, Apiezon-coated capillary column (200,000 plates) that could separate methyl stearate and methyl oleate and even methyl oleate, and methyl elaidate.

## Sensory Evaluation in the 1940s

Taste testing in the 1930s differed from that in the 1940s vegetable oil industry. In general, one man, "the expert," did the tasting of the oil for a whole company. His influence was enormous, affecting every stage of oil processing from seed storage through extracting, refining, bleaching, and deodorizing. The experts who visited our newly initiated research at NRRL[1] on the flavor stability problem of soybean oil gave valuable guidance on defining reversion flavor. The NRRL program is in their debt; however, by statistical evaluation later, these experts were found to be as vul-

[1]NRRL, the acronym for the Northern Regional Research Laboratory, United States Department of Agriculture, Peoria, IL, is so named in embossed words above its stainless steel entrance, and has been followed by acronyms NRRC and NCAUR.

nerable as taste panel members to random error. In 1945, a taste panel on oils was operating at NRRL under the physical conditions shown (Figure 1.3). Earlier, as a colleague of Mildred Boggs and panel member at the Western Regional Research Laboratory (WRRL), I had learned the basics of panel operations (24), including an introduction to A. Student's "Statistical t test."

I pause here to note that during the 1940s "statistics" was a new idea to most chemists. The teaching of physical chemist Professor Farrington Daniels in 1939 was "If your data is so bad you need to use statistical methods, you better go back to the lab and sharpen up your tools." Fortunately for him, he was not dealing with uncontrollable biological variation, but his prejudice was typical of the day. In 1906, Gosset was working in an Irish brewery that discouraged employee publication of scientific data and was forced to sign his historic paper about the "T" test under the pseudonym "A. Student" (25).

As seen in Figure 1.3, the taste panel at NRRL in the 1940s was conducted in the back of the auditorium away from laboratory odors using tables with collapsible partitions, roll-in carts with aluminum heating blocks (60°C), and beakers and cover glasses cleaned by firing in the glass blower's oven (26). The point is that reliable evaluations can be made even under crude physical conditions if the proper knowledge is used and care is taken. As evidence, we reproduce a chart of Milestones of Progress on the Flavor Stability Problem of Soybean Oil (Table 1.1)

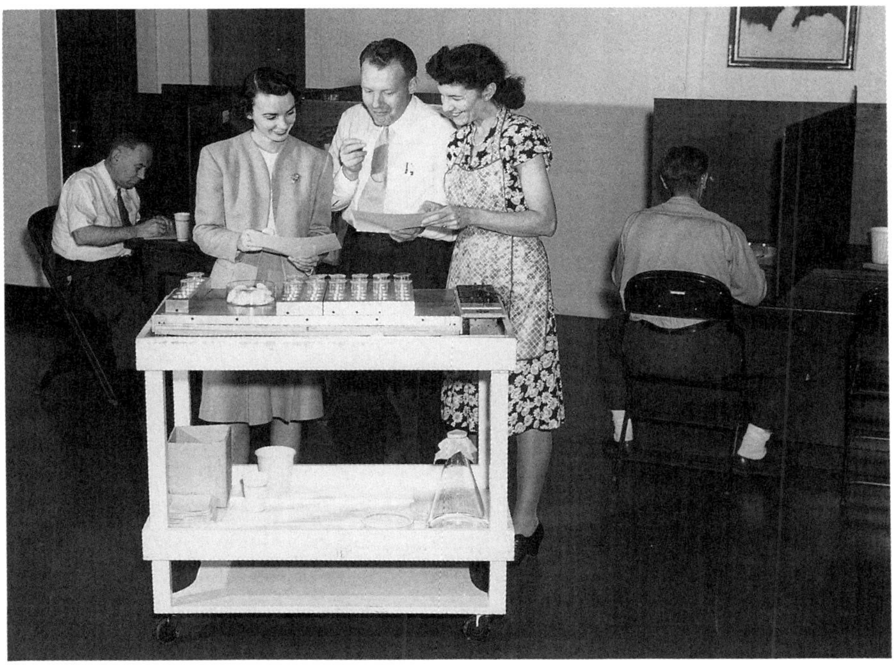

**Figure 1.3.** Early taste panel operation (1945).

**TABLE 1.1**
Milestones in Improving Flavor Stability of Soybean Oil

| Date | NRRC Research | Industry Response |
|---|---|---|
| 1945 | Standardized taste test | Worldwide acceptance |
| 1945 | Trace metals | Brass valves, Sheet steel |
| 1948 | Metal deactivators | "Nary a lb. without citric acid" |
| 1948 | Flavor is oxidation | Inert gas blanketing |
| 1951 | Precursor —linolenic | |
| | —breed it out | —Homozygous (it can't be done) |
| | —extract it out | —Practiced but now obsolete |
| | —hydrogenate it out | —"Specially processed soybean oil" |
| 1966 | Recognition of room odor problem | |
| 1966 | Copper catalysts | |
| 1974 | | Commercial production of cooking oils |
| | | —Lesieur-Cotelle, France |
| | | —Safeway, Texas |
| | | —Unilever, Holland |

(27). Each milestone of achievement on this table was made possible by the sensory evaluation procedures of the NRRL taste panel. The taste panel members shown in Figure 1.3 are in addition to Helen Moser, panel leader (right), chemist Dolores Rayleigh, who became Mrs. Karl Zilch, wife of AOCS President in 1950, and chemical engineer Warren Goss, who was to become Major Warren Goss of General Patton's staff in World War II.

Warren Goss was commissioned toward the end of the war with the assignment to follow in the wake of Patton's advancing tanks to learn the secrets of the German oilseed industry, particularly with regard to the purported solutions of the of soybean oil flavor problem. He kept hearing rumors of a recipe for solving the problem, but it was not until the troops reached Hamburg that he obtained particulars. He learned of a Dr. Tassusky and his daughter Ilona who had a patented recipe involving multiple washes of crude soybean oil with water or with water glass (sodium silicate) and adding 0.01% of citric acid to the deodorizer (28,29). When the formula was tried at NRRL, the taste panel said it was effective and highly significant. The process worked, not because of the washings, but because of citric acid addition (30). As detailed among the achievement milestones (27), we found it worked because the citric acid complexed the prooxidant metals (iron and copper) in soybean oil in the ppm and ppb range and slowed the rate of oxidation by these prooxidant metals. Suffice to say that the success of research in 1947 on "The Number One Problem of the Soybean Industry" was made possible by the results from the statistically controlled analytical taste panel. Further, I am not aware of a single observation or conclusion based on this taste panel data that has ever been recanted.

While "the proof of the pudding is in the eating," for example, sensory evaluation, chemical analyses were also sought and used as objective indices of oil quality and stability. Among the myths, rumors, and opinions circulating concerning soybean oil in the 1940s were that it reverted in flavor after being refined and deodorized; that it was extremely light sensitive, reverting if one carried the freshly deodorized oil past the light of a north window, and that the reversion was not an oxidative phenomenon. As we now know, soybean oil does not revert to its original crude oil flavor; the effect of light is real but was greatly exaggerated; the off-flavor development is indeed an oxidative reaction. In the 1940s, like today, peroxide development was the most common measure of oxidation, but under the procedure then in use, titrating with 0.1 N thiosulfate, one drop represented a PV of 6 and was dismissed as a blank reading. When we "sharpened up our tools" increasing sensitivity, we found that the flavor score was inversely correlated with PV and the prior fallacy could be explained that by the time soybean oil had reached PV 3, it had already passed through the acceptable ranges of flavor (31). With the support of the more sensitive peroxide method, the old Schaal Oven Test of simply storing the oil in the dark at 60°C for several days and tasting, became a more valuable predictor of stability. All samples presented for taste panel evaluation were routinely analyzed for PV before and after storage at 60°C.

A host of analytical procedures and equipment now stem from the relation of oxygen absorption and flavor (Chapter 2). The spectrophotometrically determined diene conjugation of linoleic and linolenic acids, the volume of oxygen absorbed by the oil, color of Kreis thiobarbituric acid color reactions, and other tests assess some phase of oxidation and correlate with off-flavor development.

Perhaps the most widely used predictor of stability is that generated from the Swift Stability Test mentioned previously. Its application to vegetable oils rather than lards immediately met a seemingly unsurmountable problem—vegetable oils and lards oxidize differently in this test. Lards hold for hours with no apparent peroxide development, then at a break point in time, indicative of the lard's stability, they develop high peroxides of 100–300 in an hour's time. Unfortunately, the PV versus time curves for vegetable oils, including soybean oil, rise gradually, but hyperbolically, from the beginning and at a constantly increasing rate and display no break with time. However, the rates of oxidation for these vegetable oils did correlate with stability under the conditions of bubbling air at 100°C.

The solution to the amount of labor required to measure the PV for a single oil on an hourly basis was to set a uniform time at which to make comparisons. The pragmatic answer to the analyst's working day of 8 h plus 1 h for lunch was the "peroxide value at 8 hours." Under this regimen, as the first act in the morning, the analyst removed samples, already in their aeration tubes from the freezer, placed them in the 100°C bath; and connected air bubbler tubes. During the day, the next set of samples were placed in new tubes and stored in the freezer, leaving the last hour of the working day for sample titration. It was upon these practical considerations the 8 h PV AOM evolved (31).

## Computer-Ease

Each lipid scientist working in the 1960s and 1970s will recall how he/she was drawn or pushed or sneaked into the back door of the computer lab. In the hope that the history of an individual, namely myself, will reflect the experience of many others, I recount my story here. Our first computer was of the analog type, purchased in 1962 as a kit and of necessity disguised from administrators as a "Reaction Rate Simulator." It was used to model the kinetics of hydrogenation (31). With the demonstration of its utility, the computer shown in Figure 1.4 was acquired and exploited.

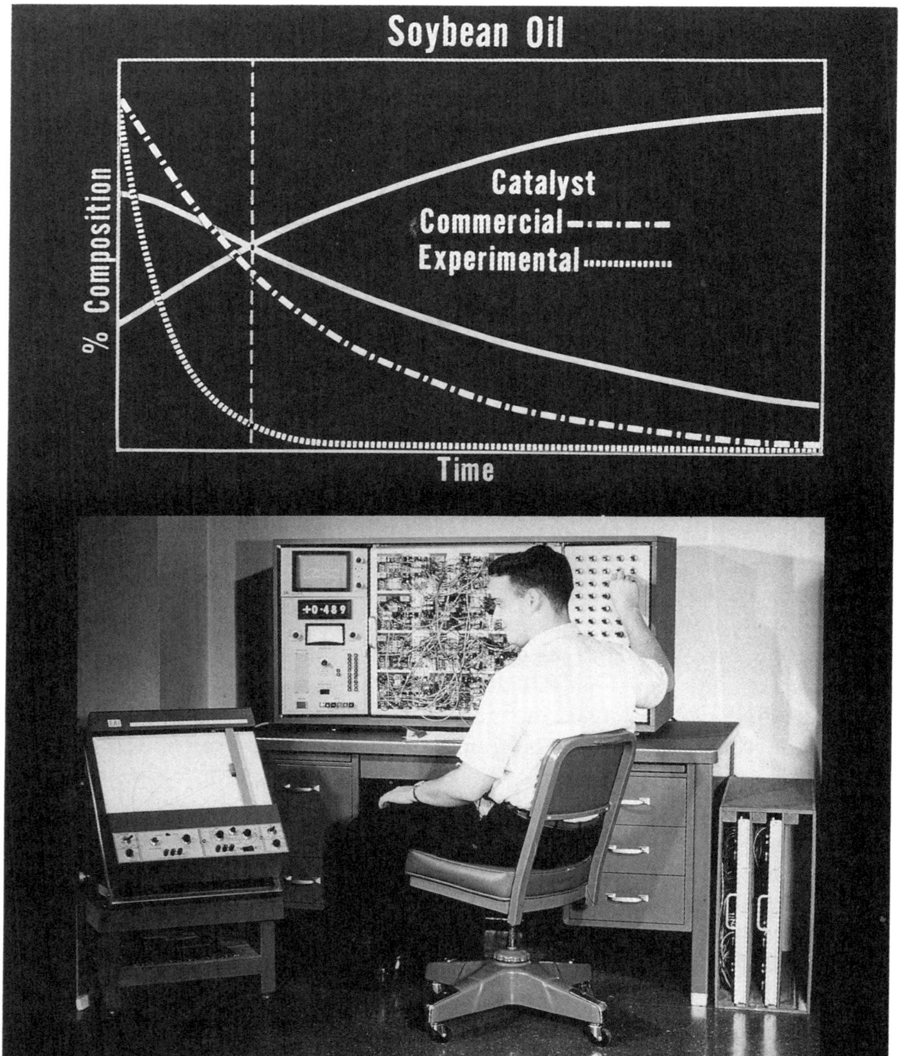


**Figure 1.4.** Analog computer used for resolving overlapping Gaussian curves and simulating kinetics of hydrogenation (32).

My introduction to the digital computer began in 1965 with my registering for a first offering of a freshman engineering course in Fortran Programming at Bradley University. A short time into this study, I saw great opportunities for research and began bootlegging early morning computer time at the University. My first project was to try to describe the countercurrent distribution process mathematically, a mathematical simulation as it came to be known. Suffice to say, I miraculously and fortunately had an operational program just two days before the end of the semester (33); most important was the confidence I now had that I and the NRRL could make effective use of this new computer tool. Back at the laboratory, I made the discovery that of the over 500 employees at NRRL; only I had computer experience and could program in Fortran. I will not relate the teaching of Fortran to the NRRL staff, single-handedly writing feasibility studies, and proposal refusals year after year. Finally, we acquired an IBM 3101 batch type computer. The memory of this advanced design had the astronomical size of 3.5 K!

By 1980, an efficient computer staff at NRRL of only three people, Chemists Roy Butterfield and programmer Darhal Wolf, were operating a centralized system in which the then expensive core, printers, and disk drives were shared by all users. Each chemist had a control unit in his lab which made the shared computer appear as his very own. At that time, we had over 60 gas chromatographs, 6 mass spectrometers, 2 spectrophotometers, a soybean-mini refinery; and terminals in the stockroom and business office on line to an IBM 1800 (or upgrade) with conduits going to 4 levels of the building and 3 wings. This installation at NRRL succeeded where other contemporary ventures failed because, I think, of a basic philosophy that computer usage was not just for the mathematical elite, but of the chemist, by the chemist, and for the chemist. Secondly, we planned from the beginning to automate the whole laboratory as opposed to computerizing individual instruments.

The interesting epilogue to this story, however, is that because of technical developments, such as visual monitors, keyboard input, the lower cost of memory, disc storage, and printers, this computer installation would probably follow a different course of development if initiated today. The conflicts of the "itty bitty binaries" vs. the "Big Blue Monster," and of personal computers vs. mainframes seem to have been solved today by their marriage.

## Hyphenation

This is a buzz word to describe a process already long underway, that of a tandem arrangement of previously separated methodologies to give symbiotic advantages. Complete sessions of analytical symposia were and are (34) devoted to hyphenation. Thus, a high-temperature ionization chamber was placed tandem to a gas chromatograph so that the specific activity of fatty acid esters could be measured as compounds being eluted. It was called "Chromatography-Radioactivity" (35). Then followed gas chromatography-infrared (GC-IR) (34), countercurrent distribution-monitoring refractometer (36), gas chromatography-mass spectroscopy (GC-MS).

Two separate papers have credited NRRL with being the first to connect GC to MS (37,38). The now ubiquitous GC-MS arrangement illustrates the complementary advantage of hyphenation, for example, while the GC is quantitatively telling us how many compounds and how much, the mass spectrometer is qualitatively telling us the identity of the peaks.

With the advent of HPLC a series of similar hyphenations, for example, -radioactivity, -infrared, -mass spectrometry, -fluorescence, were being made. Early on, the hyphenation with computers was mentioned. Now if a computer is needed to control equipment, record and store data, calculate and present graphics, it is done without fanfare or acknowledgment.

## Integration of Methodologies

Modern research knows none of the conventional disciplinary lines that are used for teaching purposes. For example, the solution of a nutritional problem may call on advanced techniques in chemistry, biochemistry, physics and mathematics indiscriminately for its successful implementation. A recent (39) and an older (27) paper are selected for this final snapshot to illustrate the integration of most of the methodologies listed previously; their significance has been demonstrated by italicizing them in the following text.

To study the metabolic role of positional isomers of fatty acids created by the hydrogenation process, rats were used as models. Three groups were fed isocaloric rations including a corn oil diet (CO), an essential fatty acid deficient diet (EFAD), and a partially hydrogenated soybean oil (PHSO) diet containing a variety of isomeric fatty acids (39). At 10 weeks, when deficiency symptoms were apparent in the EFAD group, liver phospholipids were isolated by *TLC;* converted to methyl esters; and the monoenoates, dienoates, trienoates, and tetraenoates separated by *$AgNO_3$ TLC*. The $C_{18}$ dienes and $C_{20}$ dienes form a single band on $AgNO_3$-TLC but were readily separated on a reversed phase *HPLC* column with a *tandem refractive index detector.*

The organic chemical reaction of reductive ozonolysis was used to locate the positions of double bonds in these dienoic fatty acids. Reduction of the ozonides with triphenyl phosphene yields aldehyde ester (AE), dialdehyde (AA), and aldehyde (A) fragments coming from the ester end, the middle, and the alkyl end of the dienoic fatty acids, respectively. The identification and quantitation of these aldehyde fragments of varying chain length was carried out with a *CGC-computer* system. Malonyl dialdehyde equivalent was determined independently by an injector port *alkaline-conjugation* reaction followed by *CGC* resolution of conjugated from nonconjugated diene isomers.

The fatty acid isomer composition of the individual rat liver phospholipid samples was calculated from the aldehyde analyses by a *computer solution of a matrix* with as many as twelve linear simultaneous equations. The observed mole percent of each individual aldehyde was accounted for in each equation as the sum of mole

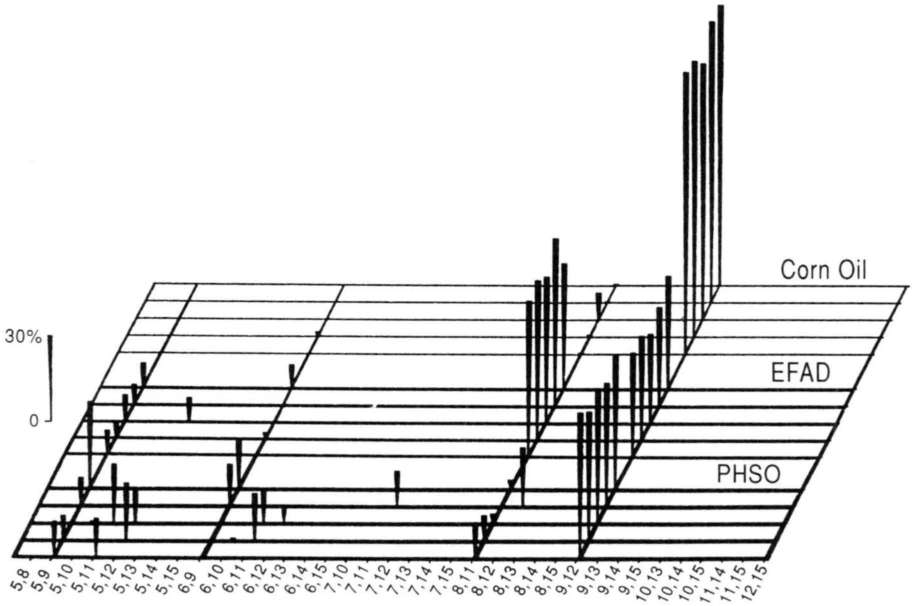


**Figure 1.5.** Isomeric 18:2 fatty acids in liver phospholipids identified and measured by computer solution of simultaneous equations using data from quantified ozonolysis of each of five samples from each dietary group expressed as percent of total 18:2 (39).

percentages of those individual fatty acid isomers contributing to that aldehyde. A computer program based on the Gram-Schmid Orthogonalization procedure provided a best solution after least squares minimization of error and yielded the fatty acid isomer composition of the individual rat liver phospholipid samples (39). Unusual isomeric polyunsaturated isomers were found in the rat liver phospholipids as shown in the computer drawn graphics of Figure 1.5.

Integration of *sensory evaluation* with the physical analytical methodology has also been illustrated in a paper previously cited (27). "A Nose in the Computer Loop," Figure 1.6, relates the aromogram of a human observer to a gas chromatogram. While the chromatogram of volatiles tells us how many compounds and how much of each compound is present, the tandem *mass spectrometer* tells us what they are, and the nose, *sensory evaluation,* tells us how significant.

## Quo Vadis

A recorder of history can with great difficulty restrain him/herself from predicting the future—and the further one sees the greater one's error. But certain trends seem apparent. The lines between the pedagogic disciplines will continue to erode, and greatest progress will continue to be made at the interfaces between the previously divided areas. Thus, psychology and physiology must play an increasing role in

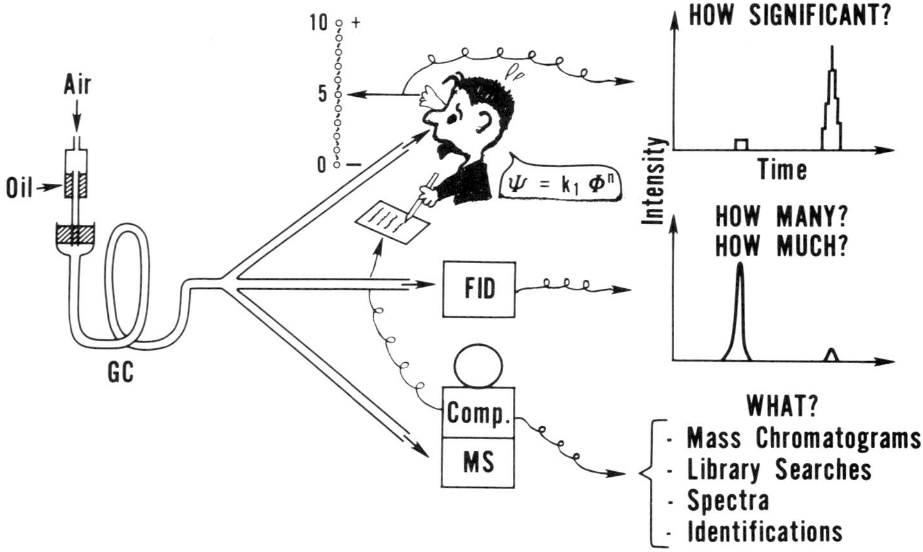


**Figure 1.6.** A nose in the GC-MS computer loop (27).

relating the physics and chemistry of sensory response. More "Nose in the Computer Loop" (Figure 1.6) experiments will evolve. Although research on an electronic nose will expand, human response will always be required at some point in time. Sensory panels must continue even though the electronic nose may well relieve the panel in certain specific routine evaluations.

The effect that shifting double bonds up and down fatty acid chains or changing *cis* to *trans* configurations during hydrogenation on oil-odor response to aging requires as much research as the role geometric and positional isomers have on metabolism. The future needs a methodology where the geometric configuration and position of double bonds on the carbon chain can be determined by using the mixtures of polyunsaturated fatty acids that occur in partially hydrogenated liquid oils, margarines, shortenings, and tissues of consuming animals. Regardless of the specific course lipid research will take in the future, we may be assured that "Every advance in scientific knowledge is first an advance in technique" (1).

## REFERENCES

1. Zechmeister, L., and Cholnkey, L.V. 1938. *Die Chromatographische Adsorption Methode*, 2nd edition, Julius Springer, Vienna.
2. *Official Methods and Recommended Practices of the American Oil Chemists' Society*, Fourth Edition, American Oil Chemists' Society, Champaign, IL, 1989. Method No. Cd-3a-63(87) Acid Value.
3. *Official Methods and Recommended Practices of the American Oil Chemists' Society*, Fourth Edition, American Oil Chemists' Society, Champaign, IL, 1989. Method No. Ca-5b-71 Free Fatty Acids.

4. *Official Methods and Recommended Practices of the American Oil Chemists' Society*, Fourth Edition, American Oil Chemists' Society, Champaign, IL, 1989. Method No. Ba-4b-57(87) Protein.
5. *Official Methods and Recommended Practices of the American Oil Chemists' Society*, Fourth Edition, American Oil Chemists' Society, Champaign, IL, 1989. Method No. Cd-8-53(86) Peroxide Value.
6. *Official Methods and Recommended Practices of the American Oil Chemists' Society*, Fourth Edition, American Oil Chemists' Society, Champaign, IL, 1989. Method No. Cd-12-57(81) Fat Stability.
7. *Official Methods and Recommended Practices of the American Oil Chemists' Society*, Fourth Edition, American Oil Chemists' Society, Champaign, IL, 1989. Method No. Cc-7-25(73) Refractive Index.
8. *Official Methods and Recommended Practices of the American Oil Chemists' Society*, Fourth Edition, American Oil Chemists' Society, Champaign, IL, 1989. Method No. Ca-1-25(88) Iodine Value.
9. *Official Methods and Recommended Practices of the American Oil Chemists' Society*, Fourth Edition, American Oil Chemists' Society, Champaign, IL, 1989. Method No. Cd-7-58(73) Polyunsaturated Acids.
10. Dutton, H.J. 1944. *J. Phys. Chem.* 48, 179.
11. Rao, D. and Roseveare, W.E.1936. *Ind. Eng. Chem. Anal.* 8:72.
12. Kirschner, J.G. and Keller, G.J. 1950. *J. Am. Chem. Soc.* 72:1867.
13. Stahl, E. (Ed.) *Thin-Layer Chromatography*, Springer-Verlag, Germany, 1965.
14. Mangold, H.K. 1961. *J. Am. Oil Chem. Soc.* 38:708.
15. Mangold, H.K. in *Thin-layer Chromatography*, edited by E. Stahl, Springer-Verlag, Germany, 1965. pp. 137–186.
16. Morris, L.J. 1966. *J. Lipid Res.* 7:717.
17. Craig, L.C. and Post, O. 1949. *Anal. Chem.* 21:500.
18. Ahrends, E.H. Jr. and Craig, L.C. 1952. *J. Biol. Chem.* 195:299.
19. Hilditch, J.P. *The Chemical Constitution of Natural Fats*, 3rd edition, John Wiley and Sons, New York, 1956.
20. Dutton, H.J. 1972. *Chem. Ind.* 17:665.
21. Butterfield, R.O., Dutton, H.J., and Scholfield, C.R. 1966. *Anal. Chem.* 38:86.
22. Staff "Key $C_{18}$ Unsaturated Fatty Acids Separated", in *C & En.*, Feb. 3, 1958, p. 522.
23. Lipsky, S.R., Lovelock, J.E., and Landarone, R.A. 1959. *J. Am. Chem. Soc.* 81:1010.
24. Boggs, M., Dutton, H.J., Edwards, B.G. and Fevold, H.L. 1946. *Ind. Eng. Chem.* 38:1082.
25. Walpole, R.E. and Myers, R.H. 1989. *Probability and Statistics for Engineers and Scientists*, 4th edition, Macmillan Publishing Company, New York, 1989.
26. Moser, H.A., Jaeger, C.M., Cowan, J.C., and Dutton, H.J. 1947. *J. Am. Oil Chem. Soc.* 24:291.
27. Dutton, H.J. *ACS Symposium Series, No. 75, Lipids as a Source of Flavor*, American Chemical Society, Washington D.C., 1978, pp. 81–93.
28. Goss, W.H. 1946. *Report on Germany—Fats & Oils and Oilseeds.* Summary on investigations. Publication No. 1270. Publication Board, U.S. Department of Commerce, Washington, D.C.
29. Tausky, I. U.S. Patent No. 2,413,009, 1946.
30. Dutton, H.J., Moser, H.A., and Cowan, J.C. 1947. *J. Am. Oil Chem. Soc.* 24:261.

31. Dutton, H.J., Schwab, A.W., Moser, H.A., and Cowan, J.C. 1948. *J. Am. Oil Chem. Soc.* 25:385.
32. Butterfield, R.O., Bitner, E.D., Scholfield, C.R., and Dutton, H.J. 1964. *J. Am. Oil Chem. Soc.* 25:385.
33. Dutton, H.J., Butterfield, R.O., and Rothstein, A. 1966. *Anal. Chem.* 38:1773.
34. Paper presented at the 1992 AOCS Annual Meeting, Toronto, Canada, May 10–14, 1992. Session EE, *INFORM* 3:499.
35. Dutton, H.J., and Mounts, T.L. 1964. *J. Catal.* 3:363.
36. Butterfield, R.O., and Dutton, H.J. 1964. *Anal. Chem.* 36:903.
37. Falkner, F.C. 1977. *Biomed. Mass Spect.* 4:66.
38. Self, R. 1979. *Biomed. Mass Spect.* 6:361.
39. Holman, R.T., Pusch, F., Svingen, B., and Dutton, H.J. 1992. *Proc. Natl. Acad. Sci. USA* 88:4830.

Chapter 2

# Factors Affecting Oil Quality and Stability

**Thomas H. Smouse**

Oil and Lecithin Process Research, Archer Daniels Midland Company, Decatur, IL USA.

## Quality and Stability Importance

Oil quality is the present state of oil acceptability, while its stability is its resistance to future changes. These characteristics can be physical, such as color, viscosity, or crystal structure, as well as chemical, such as hydrolysis, oxidation, flavor, or polymerization. In the last 20 years, with interest in the nutritional value of fats and oils on the rise and their recognized importance with atherosclerosis, cancer, arterial plaque, and other health aspects, more concern has been directed to the quality and stability of fats and oils. For the human body, as well as other animals, fats and oils are the major caloric source of energy for sustaining life.

This chapter will review the various factors that are known to affect the quality of fats and oils as they are being processed, as well as, known factors that can affect their postproduction stability. In addition to additives, such as antioxidants, antifoam agents, emulsifiers, and crystal inhibitors that can play a direct role in quality and stability, various processes are covered that can play an indirect role, especially in the stability of the product after it leaves the production facility.

Fats and oils have been a major caloric source of the human race since its beginning. As prosperity increases and countries develop, the consumption of fats and oils increase. Today, their quality and stability have never been better but, as in most ares that can affect life and good health, I am sure improvements will be made so that in the future even better products will appear on the market.

### *Introduction*

Oil stability as defined by Webster is the resistance of oil to chemical change or to physical disintegration. Quality is a peculiar or essential character and a general term applicable to any trait or characteristic, whether individual or generic. In industry, quality is normally what a consumer expects or accepts.

Therefore, the subject of "Factors Affecting Oil Quality and Stability" can mean many things to various people. The characteristics listed in Table 2.1 are generally included to evaluate the stability of an oil. Some of these characteristics overlap with each other, while others are completely independent. For example, the flavor and oxidative stability of a fat are independent variables which often are con-

**TABLE 2.1**
Types of Oil Stability

| | | |
|---|---|---|
| Oxidative | Foam | Light |
| Flavor | Emulsion | Enzymatic |
| Color | Crystal | |
| Hydrolytic | Heat | |

fused and considered to be the same characteristic. However, an oil can show excellent oxidative stability and mediocre flavor stability. A good example of this would be the comparison of cottonseed and soybean oils (SBO). Both can be processed to have an excellent flavor with a flavor grade of at least 8.0 on a 10-point scale. When oxidation rates of these two oils are compared either by gas-liquid chromatography (GLC), Active Oxygen Method (AOM), Oxygen Stability Index (OSI), or differential scanning calorimetry (DSC), generally the SBO will show slightly better oxidative stability. However, if the flavor stability of each is compared by either a Schaal Oven at 63°C for several days or room temperature storage in the presence of normal light (75–100 ft-c/ft$^2$ or 705–940 lux), the cottonseed oil will show better flavor stability.

The color of a refined, bleached, and deodorized (RBD) oil is normally very light yellow. However, during processing, various components can affect the color stability of the finished materials and can be a major quality characteristic in votated shortenings, cream filler fats, margarine base stocks, frying fats, and even liquid salad oils. Some of the components known to affect color stability are pigments, tocopherol, metals, phospholipids, and other trace materials which must be remove during processing.

Hydrolytic stability is normally not a problem in vegetable fats. In triglycerides with shorter chain fatty acids than palmitic and stearic, such as coconut oil, palm kernel oil, and dairy fats, hydrolysis of the fatty acid from the triglyceride will produce strong off-flavors such as cheesy, goaty, and soapy flavors. Such flavors will normally not be desirable in many finished food products. During frying, fatty acids are formed by hydrolysis and oxidation causing problems with flavor, smoke point, and thermal conductivity. In addition to heat, water and oxygen can cause fatty acids to form, a food system be free of lipase to prevent enzymatic hydrolysis.

Resistance to the formation of foams is a desirable characteristic of frying fats. As frying time is increased, polar and polymer compounds are formed, and the fat will eventually foam. If proper management of the frying fat is not practiced, eventually the fat will foam out of the fryer causing burns and be a potential source of oil fires. Therefore, a frying fat should have excellent foam stability. However, in the case of a baking shortening, emulsifiers are added to the shortening to increase its foaming action, so cake volume can be regulated. In these types of applications, foam is desirable. Unquestionably, a frying fat and an emulsified cake shortening should never be mixed, since both have been processed for optimal performance in their respective applications.

Emulsion stability is an important characteristic in such foods as peanut butter, salad dressings, mayonnaise, and margarine, where a change in the emulsion can affect the texture or the mouth-feel of the product. Similar observations can be made with crystal stability. A gritty-sandy texture can result in icings made from filler fats and shortenings, in which the crystals have not been stabilized by tempering.

Heat stability is normally an important characteristic of frying fats as the fat polymerizes during frying, poor thermal conduction occurs, interfering with frying life. As polymers are formed, foaming from the release of water in the fried food occurs causing problems with proper deep fat frying operation.

Consumers like to see products that they purchase. For this reason, most products are packaged in a container that allows visual inspection. However, this in turn allows visible light to reach the product that can cause off-flavors to develop from oils with poor light stability. For example, SBO or low-erucic rapeseed oil (canola) in the presence of light will develop what has been termed reversion flavor. This is a green, grassy, weedy, hay-like flavor in its early stages of development which later changes to melon, fishy, and painty flavors. Although it is believed that the compounds responsible for these flavors are oxidation products of oil, many times these flavors are observed when the oxidation is undetectable or barely measurable by methods commonly used to detect oxidation. On the other hand, oils such as cottonseed, safflower, peanut, or corn will oxidize in the presence of oxygen but not develop reversion flavor in the presence of light. Therefore, if a food product such as a potato chip or corn chip is exposed to short-wavelength light, it is desirable to use a frying fat with excellent light stability.

Although at present there are many ways to affect the stability of a fat or oil, in the future new findings may lead to methods that are at present unknown. The factors or components listed in Table 2.2 affect stability and will be discussed in later sections of this chapter. Although an edible RBD oil is mostly composed of triglycerides, the other components given in Table 2.3 are sometimes present and must be removed during refining to yield an acceptable, stable product that finds many uses in the consumer market. Although many unit processes were designed to remove one type of material, when one follows the process closely, other components can be affected that can have an effect upon oil quality and stability.

**TABLE 2.2**
Factors or Components Affecting Oil Stability

| | |
|---|---|
| Phospholipids | Oil storage |
| Soaps | Deodorization time and temperature |
| Enzymes | Deodorization cool-down rates |
| Metals | Fatty acid composition |
| Antioxidants | Pigments |
| Seed storage | Light |

**TABLE 2.3**
Components of an Edible Oil

| | |
|---|---|
| Triglyceride | Pigments |
| Diglyceride | Oxidized materials |
| Monoglyceride | Waxes |
| Fatty acids | Moisture |
| Phospholipids | Metals |
| Gums | Proteins |
| Sterols | Glycolipids |

### *Oilseeds*

To produce a good quality SBO, it is important to start with sound beans. The nutritionally desirable unsaturated acids in SBO are sensitive to oxidation and polymerization, which will produce undesirable flavors as well as off-colors. Any type of damage to the soybean can result in an oil with poor quality characteristics. Lipoxygenases, phospholipases, and lipases that are normally present in soybeans are in an inactive state in the sound bean. However, frost damage, wet beans, or cracked and ruptured beans will activate these enzymes, resulting in the production of undesirable materials. Apparent bean characteristics that are known to affect oil quality are the following.

1. Frost or immature beans—An early frost or harvesting green beans will yield high levels of chlorophyll in the oil. Such oil requires more extensive bleaching, resulting in poor oxidative stability.
2. Ground damage or moldy beans—Wet beans will have higher amounts of enzyme damage, resulting in oil with a musty odor and higher levels of nonhydratable phospholipids.
3. Split beans—Ruptured beans will have active enzyme systems, resulting in oil that has higher free fatty acids, more nonhydratable phospholipids, and more oxidation by-products.

### *Processing*

In order to produce edible oils, all oilseeds undergo similar unit processes. However, not all oilseeds are treated exactly alike. For example, cottonseeds must be delinted, while soybeans are dehulled. Some seeds, such as olives, are pressed. Others, such as corn, are expelled and extracted, while soybeans are almost exclusively extracted. Nevertheless, it is important to use good quality, sound seeds if a good quality oil is to be made. If the seed has been damaged by heat, wet weather, floods, poor storage, or other damaging conditions, then the crude oil obtained from these distressed seeds will need more processing, and the finished refined, bleached and deodorized (RBD) oil normally will not have the excellent stability characteristics shown in Table 2.1.

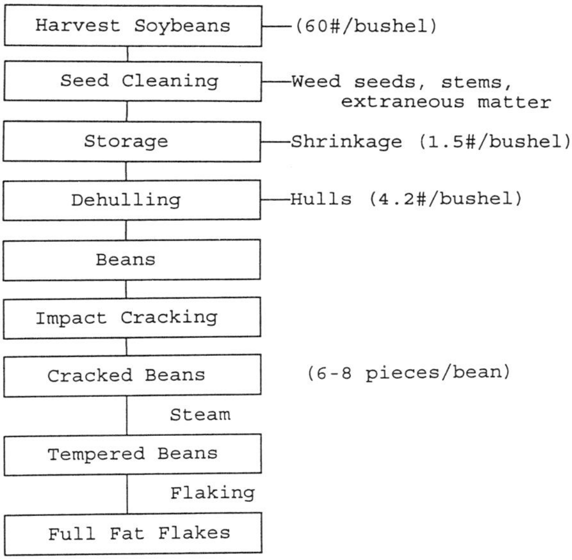


**Figure 2.1.** Initial handling of soybeans.

The treatment prior to extraction (Figure 2.1) is an example of how an oilseed is handled. For soybeans, the full fat flakes are immediately solvent extracted. Once the bean is cracked, active enzymes are released that can cause hydrolysis of fatty acids, oxidation of the unsaturated olefinic bonds, release of iron, and increase nonhydratable phospholipids. All of these changes affect quality and stability, so the flaked beans are never stored at this point and are immediately submerged in hexane. Full-fat flakes exposed to air will display surface darkening within several minutes and the oil from such flakes has been shown to have poor flavor stability.

Recently, flakes have been expanded in extruders with steam inlets which inactivate the enzymes and allow the oil to be more readily extracted by hexane. Presently, there is no published work showing that an expanded soybean oil has better stability than a nonexpanded oil. Watkins et al. (1) demonstrated that expansion enhanced the efficiency of extraction by releasing oil during the cooking process and produced a porous collet with reduced solvent retention. Although the title of this paper stated that the oil quality had improved, no supporting data was given. Anderson (2) stated that expander use has several advantages, such as higher density in the extractor, better solvent drainage, good permeability to solvent, and lower solvent content to the desolventizer/toaster (DT). He also mentioned possible improvements in oil quality, but did not elaborate. In an earlier publication, Lusas and Watkins (3) named several equipment manufacturers and discussed several advantages of incorporating expanders with solvent extraction, including cost savings. Several undesirable enzymes, such as lipase in rice bran and phospholipase in soybeans, are inactivated during expansion, yielding oil with lower free fatty acids and less nonhydratable phospholipids. Both of these results may be realized later in a finished oil with better stability characteristics.

In studies on soybeans from the same source, we compared flavor, oxidation, heat, and light stability of partially hydrogenated soybean oils made by hexane extraction, expression, and expulsion (Table 2.4). These three oils were refined, bleached, and hydrogenated to an iodine value (IV) of 100 and deodorized. The fatty acid composition showed that all oils were comparable. The extracted oil was superior in oxidative stability to either the expressed or expelled oils. The flavor stability of these three oils at 57°C also showed the extracted oil to be slightly better. All oils were comparable for light stability at room temperature and 75 ft-candles/ft$^2$ (705 lux) of cool white florescent light. However, the extracted oil had better initial heat stability at 180°C, giving 14.1 hr of OSI stability at 110°C. The expelled oil had a flat curve and still showed 2 hr of OSI stability at 110°C after 6 hr of heating at 180°C with agitation. The other two samples had little or no OSI stability after heating for 6 hr at 180°C with agitation.

**TABLE 2.4**
Comparison of Partially Hydrogenated Soybean Oil Made from Identical Soybeans

| Characteristic | Extracted Oil | Expressed Oil | Expelled Oil |
|---|---|---|---|
| Fatty Acid Comp. | | | |
| $C_{16}$ | 10.1 | 9.7 | 9.7 |
| $C_{18}$ | 4.4 | 4.5 | 4.2 |
| $C_{18:1}$ | 52.6 | 52.3 | 51.9 |
| $C_{18:2}$ | 30.3 | 30.2 | 31.2 |
| $C_{18:3}$ | 2.2 | 2.4 | 2.4 |
| Polymers (% GPC) | 0.6 | 0.5 | 0.6 |
| Iodine Value | 98.2 | 99.0 | 100.2 |
| Flavor Stability @ 57°C | | | |
| Initial | 8.0 | 8.0 | 8.0 |
| 2 days | 8.0 | 6.0 | 7.0 |
| 9 days | 8.0 | 8.0 | 7.0 |
| Light Stability @ 75fc | | | |
| Odor Evaluation[a] | | | |
| Initial | 1 | 1 | 1 |
| 1 day | 1 | 1 | 1 |
| 6 days | 1 | 2 | 1 |
| 15 days | 3 | 3 | 2 |
| Heat Stability @ 180°C | | | |
| OSI @ 110C (hrs.) | | | |
| 0 hrs. | 14.1 | 11.2 | 11.9 |
| 2 hrs. | 8.4 | 7.6 | 8.6 |
| 4 hrs. | 4.3 | 2.2 | 5.6 |
| 6 hrs. | 0.2 | — | 2.0 |

[a] 1 = No green reversion.
2 = Slight reversion.
3 = Moderate reversion.

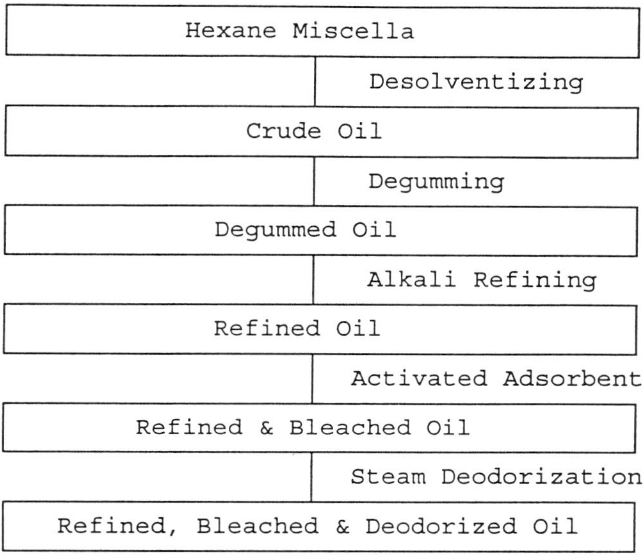


**Figure 2.2.** Standard unit processes of edible oils.

The major unit processes used to produce an RBD salad oil from a hexane-extracted oil are given in Figure 2.2. Initially, the miscella is approximately 60°C and has an oil content of about 30%. The first processing step, removing the solvent, is normally done by heating in several stages, the first effect, second effect, and steam stripping, to reduce any residual hexane in the extracted crude oil to below 500 ppm. Since this oil still contains phospholipids, the heat used to recover the solvent will darken the oil. Excessive heat can burn the oil and heat-set the color, causing a dark oil with poor color stability.

The time and temperature that the crude, desolventized oil is held will cause several effects in the quality of the finished oil. First, if it is kept at too high a temperature for too long, the phospholipids will not be removed easily during the degumming step, yielding a degummed oil with a high phospholipid content. Such an oil has been shown to have poor color stability after deodorization and poor flavor stability after aging. However, phospholipids have been shown to be effective antioxidants and chelation agents and will increase the oxidative stability of a salad oil. Degummed soybean oil must have less than 200 ppm phosphorus (P). Normally, industry uses the P content multiplied by 30 to express the phospholipid content remaining in the degummed oil, and it is not unusual for an excellent water-degummed SBO to have 20–30 ppm P, which would be 0.06–0.09% phospholipids. A more precise factor, in the range of 24–26, (4,5) has been demonstrated by analyzing the phospholipid composition.

The more the crude oil is cooled, the longer it can be stored and still possess excellent water-degumming properties. Crude SBO that could be initially water degummed to less than 50 ppm P, could not be degummed to less than 200 ppm P when aged for 7 months at 13°C (Smouse, unpublished work). Common crude oil temperatures going to storage tanks are around 60°C, and oils are seldom held longer than a month. Therefore, a crude oil held at 60°C with stirring showed poor degumming results upon reaching a peroxide value (PV) of 20, or a conjugated diene value (CD) of 0.50.

By aging the crude oil at 93.5, 76.5, 60, 38, and 10°C and measuring the PV and CD, nonlinear regression equations can be obtained predicting the aging time before degumming problems occur. Figure 2.3 shows the PV relationship with time at five different temperatures and Figure 2.4 shows the CD relationship for these same oils with storage time. By using values of 20 PV and 0.5 CD where degumming was found to be less than 50 ppm P, one can see at 60°C storage that crude soybean oil should be degummed before 15 days of storage (Table 2.5). At higher storage temperatures, such as 76.5°C, the oil cannot be degummed well after a storage time of several days.

Recent data (6) showed that the storage time of crude oil at 60°C will affect the tocopherol and the phosphatidylethanolamine contents. The tocopherols are oxidized to a chroman-5,6-quinone (tocored), which has poor antioxidant properties and is a deep red color. In 1989, oxidized tocopherols were shown to be responsible for the color reversion of soybean salad oil (7). The chroman-5,6-quinone has also been shown to cause color reversion in cottonseed oil (8) and in corn oil (9). Therefore, the contents of γ-tocopherol or γ-TED (5-tocopheryloxy)-γ-tocopherol in crude SBO can be used as an index for predicting color quality of the soybean salad oil. If crude SBO contains more than 320 ppm γ-tocopherol, 160 ppm of γ-TED, or 480 ppm of both components, the finished soybean oil will meet the color requirements of the National Soybean Processing Association (NSPA).

Most crude oil is degummed to prevent foots from forming and settling in storage facilities. The primary purpose of degumming is to remove the phospholipids.

**TABLE 2.5**
Water Degumming Two Soybean Oils

| Days | PV | | CD | | Phosphorus | |
|---|---|---|---|---|---|---|
| Storage Time @60°C | A | B | A | B | A | B |
| 0 | 0.6 | 0.8 | 0.18 | 0.19 | 47 | 30 |
| 2 | 1.3 | — | 0.17 | — | 44 | — |
| 6 | 2.5 | 2.1 | 0.20 | 0.21 | 47 | 30 |
| 10 | 3.5 | 5.0 | 0.23 | — | 38 | — |
| 13 | 4.6 | — | 0.22 | 0.26 | 44 | 24 |
| 15 | 6.6 | — | 0.28 | — | 44 | — |
| 16 | 8.8 | 8.8 | 0.28 | 0.31 | 34 | 27 |
| 20 | 22.5 | 13.2 | 0.52 | 0.40 | 38 | 19 |
| 23 | 38.6 | 19.3 | 0.69 | 0.47 | 51 | 31 |
| 27 | 61.0 | 28.0 | 0.96 | 0.59 | 103 | 77 |
| 33 | — | 33.0 | — | 0.70 | — | 130 |
| 41 | — | 40.0 | — | 0.82 | — | 193 |

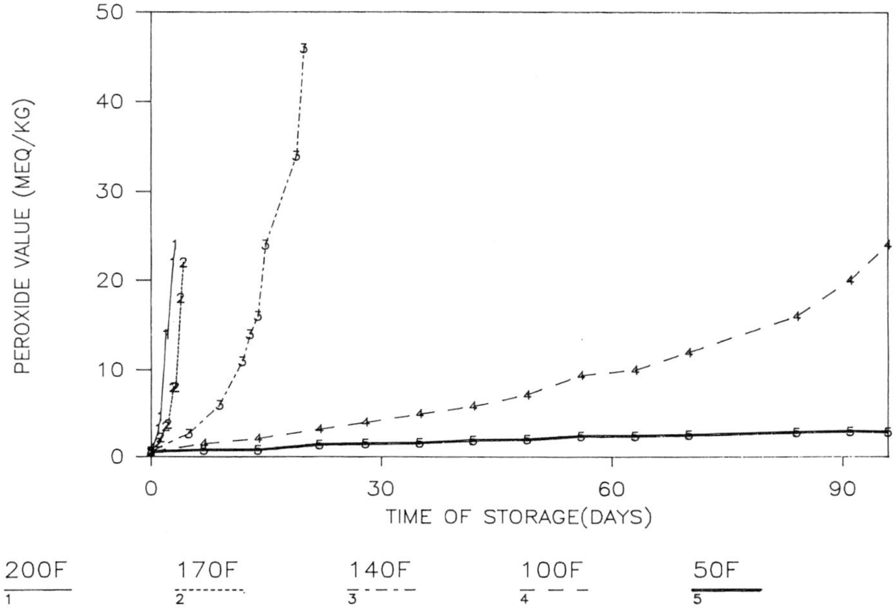


**Figure 2.3.** Crude soybean oil stability, peroxide formation vs. time.

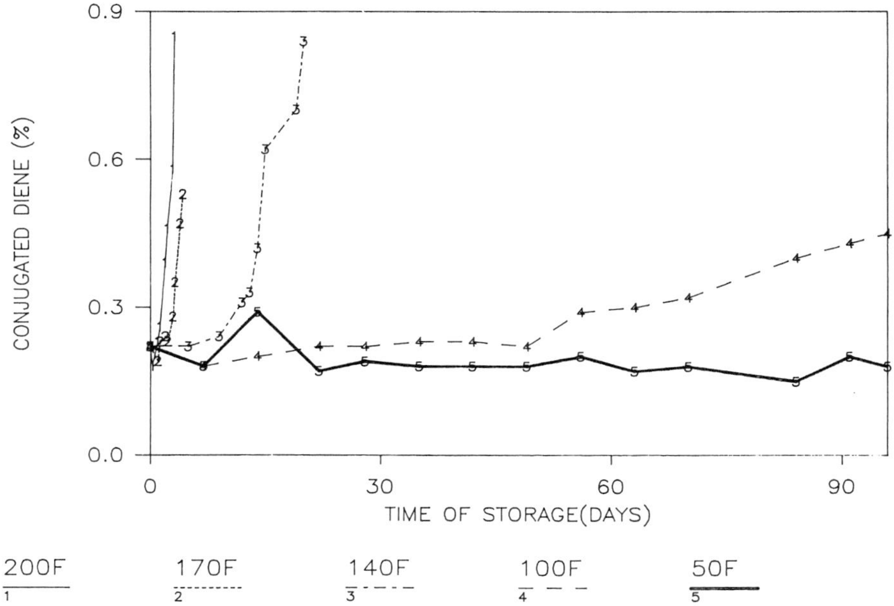


**Figure 2.4.** Crude soybean oil stability, conjugated diene formation vs. time.

In addition to the age of the crude oil and the storage temperature, the water quality used can affect the stability of the finished oil. In a study of degumming crude SBO with deionized/distilled water versus water containing $CaCO_3$, $MgCO_3$, $FeCl_2$, and NaCl, it was found the stability of the degummed oils decreased as salt concentrations increased (10). The oxidation rate was greater for oils degummed in the presence of $FeCl_2$ than with NaCl, $CaCO_3$, or $MgCO_3$ under the same conditions. Also, when phosphoric acid is used for acid degumming, melon flavors form in aged SBOs, while only the typical green reversion flavor occurs when citric acid is used (11).

List et al. (12,13) showed that the quality of a finished SBO processed with phosphoric acid and physically refined was equal to or greater than the same oil which was water degummed and steam refined. Both the initial flavor intensity score and the 8-hr PV at 97.8°C AOM conditions for the phosphoric acid-treated sample were superior. However, light stability at 4 hr under florescent light of the caustic-refined oil was better. For these reasons, it is common practice to use steam condensates or deionized water when water degumming. It is well known that the use of phosphoric acid in degumming causes the recovered lecithin to be a darker green color than water degumming (13), so it is not used when degumming is done for edible lecithin production.

Typical degumming efficiencies will reduce the phospholipids from about 3% in crude oil to less than 0.09% in degummed oil, a 97% reduction. Any remaining phospholipids in the degummed oil are removed during caustic refining. In the case of physical refining, an acid-pretreatment step is necessary to reduce phosphorus from phospholipids to less than 3 ppm. Beal et al. (14) showed that phosphorus in a water-degummed oil should be in the range of 2–20 ppm for optimal oxidative stability. It has been suggested (15) that SBO for physical refining have a maximum of 200 ppm P. However, work conducted in the author's laboratory (Smouse, unpublished) as well as work published by Beal et al. (14), showed that P levels should not be greater than 20 ppm for color stability during deodorization.

Refined and bleached SBOs were selected having the phospholipid contents shown in Table 2.6. Each RB SBO was deodorized in an all-glass batch laboratory apparatus for 1.5 hr at 235°C with 3.6% steam. The color values are Hunter tris-

**TABLE 2.6**
Relationship between Phospholipids and Deodorized Oil Color

| Sample | Phosphorus (ppm) | Phospholipids[a] (ppm) | RBDO Color Values $L$ | $a$ | $b$ | $\Delta EW$ |
|---|---|---|---|---|---|---|
| 1 | <4 | <120 | 56.6 | −1.9 | 8.1 | 36.6 |
| 2 | 6 | 180 | 53.7 | −2.5 | 13.5 | 40.7 |
| 3 | 19 | 570 | 45.7 | 0.8 | 20.1 | 50.4 |
| 4 | 34 | 1020 | 39.9 | 3.9 | 21.6 | 56.6 |
| 5 | 54 | 1620 | 36.3 | 6.4 | 21.7 | 60.2 |
| 6 | 66 | 1980 | 33.0 | 8.0 | 20.8 | 63.3 |

[a] ppm P × 30

timulus values, or *L* (lightness), *a* (redness), *b* (yellowness), and $\Delta EW$ (total color difference from a white standard plate). The sample containing 19 ppm P was significantly darker, more yellow, and had a greater total color difference than the 6 ppm P sample.

Trading specifications for an RBD SBO is a maximum of 2.0 red on the Lovibond scale. This is comparable to a $\Delta EW$ of 50 on the Hunter scale. Therefore, the phospholipids in an RBO should be less than 20 ppm P for excellent color stability as well as oxidative stability. An evaluation of RBD SBO salad oils from six suppliers (Table 2.7) showed that P was normally below 3 ppm in commercial samples.

## Alkali Refining

Some SBO is physically refined, but when the phosphatides in the degummed oil are too high and darkening occurs during the physical deacidulation process, the oil must be alkali refined. This is also known as chemical refining, since sodium hydroxide is commonly used to remove the free fatty acids as soap. Most SBO in the United States is chemically refined, although much interest and research has been invested in physical refining. Normally, oils low in phospholipids, such as coconut, palm kernel, or palm oils, are physically refined.

The main purpose of refining is to remove free fatty acids. Mistry and Min (16) showed that free fatty acids acted as prooxidants, so they should be removed for good oxidative stability. A review and original work presented by Min and Jung (17) showed the effect free fatty acids have upon gas chromatographic headspace volatiles and hydroperoxide development. It is believed that the free carboxyl group of the fatty acid is responsible for the prooxidant activity of free fatty acids.

**TABLE 2.7**
Typical Values of Commercial RBD Soybean Oils from Six Suppliers

| Sample | Supplier | P (ppm) | Fe (ppm) | FFA as Oleic | AOM (PV@ 8 hr) |
|---|---|---|---|---|---|
| 1 | A | 3 | 0.2 | 0.03 | 381 |
| 2 | A | 3 | 0.1 | 0.03 | 362 |
| 3 | A | 3 | <0.1 | 0.02 | 436 |
| 4 | B | 5 | <0.1 | 0.03 | 26 |
| 5 | C | 3 | <0.1 | 0.06 | 298 |
| 6 | C | 3 | <0.1 | 0.05 | 116 |
| 7 | C | 3 | <0.1 | 0.03 | 60 |
| 8 | C | 4 | <0.1 | 0.04 | 157 |
| 9 | D | 3 | <0.1 | 0.05 | 122 |
| 10 | D | 3 | <0.1 | 0.03 | 178 |
| 11 | D | 4 | 0.1 | 0.04 | 53 |
| 12 | E | 4 | <0.1 | 0.03 | 84 |
| 13 | E | 3 | 0.1 | 0.03 | 51 |
| 14 | F | 4 | — | 0.04 | 11 |
| Average | | 3 | <0.1 | 0.04 | 167 |
| SD | | 0.6 | — | 0.01 | 143 |

**TABLE 2.8**
Effect on Metals from Caustic Refining of Soybean Oil

| | Metals in Degummed Oil (ppm) | | | | Metals in Refined Oil (ppm) | | | |
|---|---|---|---|---|---|---|---|---|
| Sample | P | Fe | Mg | Ca | P | Fe | Mg | Ca |
| 1 | 134 | 0.1 | 7 | 8 | 3 | <0.05 | 0.2 | 0.2 |
| 2 | 21 | 0.2 | 5 | 13 | 6 | 0.1 | 0.2 | 0.4 |
| 3 | 53 | 0.1 | 3 | 4 | 7 | <0.05 | 0.5 | 0.8 |
| 4 | 108 | 0.4 | 23 | 44 | 5 | <0.05 | 0.5 | 0.9 |
| 5 | 83 | 0.1 | 9 | 11 | 3 | <0.05 | 0.1 | 0.2 |
| 6 | 168 | 1.1 | 37 | 81 | 4 | <0.05 | 0.8 | 1.2 |
| 7 | 78 | 0.4 | 19 | 35 | 3 | <0.05 | 0.3 | 0.5 |
| 8 | 75 | 0.4 | 18 | 37 | 7 | <0.05 | 0.8 | 1.6 |
| 9 | 129 | 1.4 | 28 | 66 | 3 | <0.05 | 0.7 | 1.3 |
| 10 | 72 | 0.1 | 6 | 10 | 2 | <0.05 | 0.2 | 0.3 |
| Avg. | 92 | 0.4 | 16 | 31 | 4 | <0.05 | 0.4 | 0.7 |

In addition to removing the free fatty acids, chemical refining also reduces any phospholipids not removed by degumming. Metal soaps that are formed are also removed. Typical metal levels for SBO before and after refining are given in Table 2.8. All have been reduced 95.0–97.7% by the refining process. Iron, a very strong prooxidant, was reduced to nondetectable levels to promote excellent oxidative stability.

Although the postalkali-refining metal data did not include sodium, it is necessary to remove the soap by either water washing or silica refining. Sodium levels of 5 ppm or greater will cause a fat to have poor frying stability. Also, alkaline treatment of unconjugated, unsaturated double bonds, as in linoleic and linolenic acids, will cause positional isomerization and increase the level of conjugated dienes and trienes. These are known to have poor oxidative stability and polymerize faster due to possible Aldol condensation. Therefore, during alkali refining, a minimum amount of alkali is used, the reaction is carried out at low temperatures for as short a time as necessary, all soaps and unreacted alkali are removed for good oil stability. An alternative to caustic is Britesorb NC, a sodium silicate treatment that has been shown to remove free fatty acids while leaving soaps, metals, and phospholipids at levels low enough to eliminate water washing (18). Recent data (19) has shown that modified refining with TriSyl 300, a silica hydrogel will reduce the soaps, metals and phospholipids more effectively than water-washing alone and produce an oil with better oxidative and flavor stability.

In 1947, Goss reported on the use of silicate in oil refining (20). Chemists at several German oil mills believed traces of lecithin in SBO would cause flavor reversion, or "Umschlag" as they called it. Kugler at the Thorls Oelfabriken in Hamburg-Harburg, Mohr at the Norddentschen Olmuhlenwerke in Hamburg-Altona, and Gehrke at the Noblee und Thorl plant in Hamburg, all believed that during refining, any trace of lecithin must be removed. Some of these plants were using sodium silicate or waterglass for the alkali in caustic refining.

## *Bleaching*

The bleaching process is primarily performed to reduce color. Webster defines the word "bleach" as "to remove color or stains, or to make whiter or lighter." In chemical-bleaching processes where oxidants, such as peroxides or hypochlorites, are used, a lighter or whiter product is obtained. However, in oil bleaching, adsorptive materials such as clays, activated earths, diatomaceous earth (DE), carbons, or various types of silicas are used to adsorb the pigment and remove it when the adsorptive material is removed from the oil.

A recent monograph (21) discussed bleaching, including adsorbents, the process itself, filtration and filters, and important tests related to bleaching. Several of the tests conducted on the bleached oil such as PV, anisidine value (AV), trace metals, and oxidative stability tests like AOM, OSI, and the Schaal Oven Test are all related to oil stability. As in refining, other important adsorption phenomena occur during bleaching in addition to the removal of pigments. For example, previous work of Cowan (22) compared a bleached and an unbleached SBO. The bleached oil had better initial flavor, better aged flavor, and oxidized much more slowly than the unbleached sample. Therefore, in addition to reducing the pigments, bleaching also improved flavor, flavor stability, and oxidative stability.

Chlorophyll is a strong prooxidant that is catalyzed by light. As a photosensitizer, it produces singlet oxygen that can initiate the oxidation of oils (23–25). Once singlet oxygen is formed, oxidation proceeds by an ene reaction forming a *trans* configuration of an unsaturated hydroperoxide. Such reactions are unaffected by antioxidants, but can be inhibited by singlet oxygen quenchers such as carotene. Crude SBO has about 1 ppm chlorophyll and its derivatives, which must be removed to make a stable, light-colored oil acceptable to the consumer.

Crude canola oil has high amounts of chlorophyll, normally, below 30 ppm, but at times it can be much higher. Although the chlorophyll is reduced by degumming and refining, the major reduction comes during bleaching. Some minor reductions have also been observed during deodorization. Many processors will bleach to a 50 ppb maximum (21) and some to even lower levels to eliminate the prooxidant effect of chlorophyll. Mag (26) stated that the concentration of chlorophylloid compounds must be reduced to 50 ppb or lower to avoid rapid oxidation of the oil in the presence of light. However, a study on total chlorophyll in seven SBO samples from three suppliers (24) showed that none were below the 50 ppb maximum level.

Since adsorption chemistry occurs during bleaching, other oil components are adsorbed. In addition to pigments, oxidation products, metals, soaps, phospholipids, and polymers can be adsorbed. A reduction in these materials will normally improve stability. However, bleaching can also form materials that will cause poor stability. For example, an acid clay used for too long at too high a temperature will cause hydrolysis and an increase in free fatty acids. As mentioned previously, these act as prooxidants. Also, heat treatment during bleaching will dismutate the hydroperoxides reducing the PV and increasing the AV. The AV is a measure of enals and dienals formed when the hydroperoxide dismutates. These secondary oxidation products are normally *trans* conjugated systems that promote oxidation. Oxidation during bleaching is normally suppressed by vacuum bleaching at pressures between 50–100 mm Hg. Hydrolysis and isomerization is normally minimized by limiting the time and temperature (27).

In addition to chlorophyll a and b other chlorophylloid compounds are pheophytins a and b, pyropheophytins a and b, and pheophorbides a and b. Endo et al. (28) compared the prooxidant activities of chlorophyll as well as their decomposition products. Chlorophyll b showed greater prooxidant activity than chlorophyll a. However, the pheophorbides and pheophytins exhibited more activity than the chlorophylls. Therefore, a comprehensive bleaching study must include more measurements than just color reduction, even though this is still the primary reason for bleaching.

Much work has been done in two-stage bleaching, where an adsorbent for one type of compound is followed by another adsorbent for a second type of compound. Silica has been shown to be very good for metals, phospholipids, and soaps. Welsh et al. (29) discussed some of the uses of amorphous synthetic silicas and their effectiveness in removing phospholipids, phosphorus, calcium, and magnesium. In two-stage bleaching, silica adsorbs and captures some of these compounds, enabling clay to be more effective in reducing pigments. For example, in canola oil pretreated with silica, the clay volume required to reach 50 ppb chlorophyll was reduced by 50% in batch bleachings. Batch bleachings with clay followed by press bleaching with silica reduced the clay usage even further.

List and Erickson (30) reviewed the effects of iron and copper in crude and fully refined SBO. For excellent oxidative stability, an RBD oil should have less than 0.1 ppm Fe and less than 0.02 ppm Cu. Data in Table 2.9 on metal concentrations for 10 commercial SBO samples before and after bleaching, demonstrate the tremendous effect bleaching has on reducing metals.

### *Deodorization*

Normally, deodorization is the last unit process in salad oil production other than final polishing filtration. In this process, most of the volatile components in an oil are removed by the high temperature, high vacuum, and countercurrent steam distillation. In addition to removing volatile flavor components, deodorization also removes residual solvents; reduces fatty acids to very low levels; dismutates and removes hydroperoxides as volatiles; strips pesticides and reduces relatively nonvolatile components such as tocopherols, sterols, and mono- and diglycerides.

Color stability at high heat, such as during frying, has been related to the tocopherols, 5-(tocopheryloxy)-$\gamma$-tocopherol, and chroman-5,6-quinone. Klagge and Sen Gupta (31) showed that deodorization temperatures of 235°C and above, stripped substantial amounts of tocopherols from the oil which collected in the deodorizer condensates. This produced a deodorizer distillate with higher levels of tocopherols which industries can remove and purify for concentrates with antioxi-

**TABLE 2.9**
Typical Metal Levels in Soybean Oil Before and After Bleaching

| | Metals in RSBO (ppm) | | | | Metals in RB SBO (ppm) | | | |
|---|---|---|---|---|---|---|---|---|
| Sample | P | Fe | Mg | Ca | P | Fe | Mg | Ca |
| 1 | 0.4 | <0.05 | 0.2 | 0.3 | <0.5 | <0.05 | 0.1 | 0.2 |
| 2 | >.2 | <0.05 | 0.5 | 0.8 | <0.5 | <0.05 | <0.05 | <0.05 |
| 3 | 6.2 | 0.1 | 0.2 | 0.4 | <0.5 | 0.05 | <0.05 | 0.05 |
| 4 | 4.6 | <0.05 | 0.5 | 0.9 | 1.6 | <0.05 | 0.1 | 0.2 |
| 5 | <0.5 | <0.05 | 0.2 | 0.2 | <0.5 | <0.05 | 0.1 | 0.2 |
| 6 | 2.6 | <0.05 | 0.1 | 0.2 | <0.5 | <0.05 | <0.05 | 0.05 |
| 7 | 3.0 | <0.05 | 0.2 | 0.2 | <0.5 | 0.05 | <0.05 | <0.05 |
| 8 | 3.8 | <0.05 | 0.8 | 1.2 | <0.5 | <0.05 | 0.1 | 0.1 |
| 9 | <0.5 | <0.05 | 0.1 | 0.1 | <0.5 | <0.05 | <0.05 | 0.06 |
| 10 | 6.1 | 0.1 | 0.1 | 0.3 | <0.5 | 0.05 | <0.05 | <0.05 |
| Avg. | 3.5 | <0.05 | 0.3 | 0.5 | <0.5 | <0.05 | <0.05 | 0.1 |

dant activity. However, prior to this work, Scavone and Braun (32) showed that tocopherols could be stripped from oil in less than 15 min. at temperatures between 277°C and 343°C, and vacuum of 0.5 mm Hg and steam with a molar ratio of stripping medium to oil of about 0.05–9.7. Vacuum-steam stripping at very high temperatures for short residence times improved the frying life of the edible oil, and minimized undesirable thermally induced side reactions, such as polymerization, *trans*-isomerization, and hydrolysis. Their fry life was defined as the amount of time it took for a frying oil to darken in color to an absorbance of 1.4 @ 520 nm after deep-frying foods. This is a darkening color end point beyond which the oil quality is judged to no longer produce acceptable fried foods. Although the color stability of such a treated oil was found to be better, removing the natural antioxidants by this process produced an oil with poorer oxidative stability.

High deodorization temperatures can also abuse the oil by causing positional *trans*-isomerization and increasing the levels of conjugated dienes and trienes. Such reactions are time/temperature dependent, thus at higher temperatures only short times should be used. The time and temperature of deodorization can affect initial flavor, flavor stability, and oxidation stability (Smouse, unpublished work). The unit used in this study was a Votator pilot model, single tray, black iron deodorizer operated at 6 mm Hg pressure and 3–4% steam strip. Various quality characteristics of the deodorized SBO are given in Table 2.10. From this series of runs, the 1-hr deodorization at 245°C gave a low CD, excellent initial flavor, and good 3-day flavor stability at 60°C. The same oil deodorized at 261°C had poorer flavor stability and a higher CD content. The CD content was greater at 265°C, than that observed at 225°C, even though the time was only one-third as long.

Thermal polymerization also can occur during deodorization. Chang and Kummerow (33–37) showed that polymers from oxidized ethyl linolenate can cause off-flavors and poor flavor stability. These polymers were found to decompose under nitrogen, yielding flavor compounds identical to those isolated from reverted

**TABLE 2.10**
Stability Characteristics and their Relationship to Deodorizing Temperatures

| | Sample | | | | |
|---|---|---|---|---|---|
| Runs | 1 | 2 | 3 | 4 | 5 |
| Deod. temp. (°C) | 225 | 245 | 261 | 261 | 265 |
| Deod. time (hr.) | 1 | 1 | 1 | 0.33 | 0.33 |
| Steam strip (%) | 3 | 3 | 3 | 3 | 3 |
| Conjugated diene (%) | 0.33 | 0.46 | 0.71 | 0.46 | 0.53 |
| Initial flavor | 8.0 | 8.7 | 8.4 | 8.7 | 7.5 |
| Aged flavor (3 days at 60°C) | 6.8 | 8.1 | 7.3 | 8.1 | 5.8 |
| Viscosity (cS at 77°F) | 34.1 | 34.1 | 35.5 | 35.1 | 36.4 |

SBO. Holm et al. (38) showed that high molecular weight compounds isolated from oxidized rapeseed oil had no distinctive flavor, but intense flavors developed when they were heated.

In addition to deodorization time and temperature affecting oil flavor, the cool-down rate after deodorization also can play an important part in the final oil flavor quality. For this reason, it is good practice to cool the oil under vacuum and sparge steam before it leaves the deodorizer and enters the cooling heat exchanger.

## Other Stability Effects

### *Inert Gases*

Since all vegetable oils contain fatty acids that can react with oxygen and deteriorate over time, packing under inert gases or keeping inert blankets on storage vessels will decrease the rate of oxidation. Normally, the oxygen must be reduced to 2% or less to reduce oxidation effectively (39,40). If oxygen is displaced by inert gases, a better reduction of oxidation can be obtained. For effective oxygen displacement, levels of oxygen remaining in the product should be less than 0.1% (41).

Nitrogen is normally used as the inert gas because of cost and availability. Mickel et. al. (42) showed that several inert gas and nitrogen mixtures were the most effective for reducing oxidation. In this study, various breathing gas mixtures all containing similar oxygen partial pressures were evaluated by aging methyl oleate with 100% $O_2$, 79% He/21% $O_2$, 82% Ar/18% $O_2$, and 80% $N_2$/20% $O_2$. Peroxide accumulation at 50°C for the $N_2/O_2$ mixture was less than the $Ar/O_2$ mixture, which in turn was less than the $He/O_2$ mixture. Naturally, all the inert gas mixtures produced less peroxides compared to pure oxygen. It was speculated these differences were caused by a physical interaction between the gases and lipids. Relative solubilities of the inert gases in the lipid layer was not an adequate explanation.

Although inert gases inhibit oxidation, Robinson (43) found that SBO aged under inert gases developed reversion flavors, and Bickford (44) reached the same conclusion by storing SBO under a high vacuum. Therefore, small amounts of oxygen must be soluble in the oil and are not completely removed by gas displacement or high vacuum. Such small amounts could initiate oxidation by singlet oxygen formation or metal catalysis. Obviously enough oxygen was present to produce flavor compounds, even though oxidation was retarded.

## *Sequestrants*

During oil processing, efforts are made to reduce metals, especially iron and copper. Holding tanks, transfer lines, reaction vessels, rail cars, and other storage facilities are normally constructed from black iron that must be "sweetened" before use by allowing a thin film of oil to form a strong bond with the vessel. Metallic divalent ions, such as iron and copper, can reduce hydroperoxides when the ions are oxidized to the higher valence state, and they can oxidize unsaturated bonds when they are reduced to the lower state. As mentioned earlier, both metals are strong prooxidants, but copper is much stronger than iron. Therefore, any copper piping or brass fittings and valves must be eliminated throughout the entire refinery if good quality oil is expected.

Additives such as chelators or sequestrants are normally added to improve quality by binding trace amounts of metals that were not removed during processing. Even with such modern and sophisticated instrumentation as induced coupled plasma emission spectrophotometers that can detect hundredths of a ppm, additives are still used. Citric acid is the most widely used, but others, such as phosphoric acid, phospholipids, and phosphates, can be used in various applications. Since oils are hydrophobic and most sequesterants are hydrophilic, it is difficult to get a stable mixture of the two. Excess of citric acid is used, but later can be found on polishing filters before the oil is shipped. Excellent publications on sequestrants (45-47) show multiple uses and multiple compounds. Elimination of the prooxidant catalytic effect of metals is needed in edible oils, inedible fats, salad dressings, spreads, roasted nuts, fried foods, baked products, essential oils, margarines, and other lipid-containing foods.

## *Antioxidants*

Synthetic as well as natural antioxidants are used to reduce and slow the rate of oxidation. Several synthetic compounds are butylated hydroxyanisole (BHA), butylated hydroxytoluene (BHT), mono-tertiary butylhydroquinone (TBHQ), propyl 3,4,5-trihydroxybenzoate (PG), nordihydroquaiaretic acid (NDGA), thiodipropionic acid (TDPA), dilauryl thiodipropionate (DLTDP) and 1,2-dihydrox-6-ethoxy-2,2,4-trimethyl quinoline, or ethoxyquin (EQ). These antioxidants terminate free radical reactions by donating a hydrogen and accepting a free radical. All have been used, but certain ones can cause quality problems. For example, PG will form a pink-red color in the presence of iron, which causes an off-color. Thiodipropionic acid and DLTDP can add a sulfur/mercaptanlike flavor and EQ only has approval status for the feed industry, although it is an excellent and efficient antioxidant. Also, NDGA was found to be toxic and was removed from the GRAS list by the FDA in 1967.

Many consumer groups and some countries feel natural antioxidants have more health-promoting qualities. Ironically, NDGA was obtained from a natural bush, *Larrea divaricata*. Houlihan and Ho (48) listed and discussed many natural antioxidants, such as phenolics from sesame; caffeic and ferulic acids from oats; ascorbic acid; certain amino acids; browning Maillard reaction products; and of course, the flavonoids from vegetables such as quercetin, myricetin, and gossypetin.

The δ-tocopherol homologue is a most effective antioxidant. However, the tocopherols alone are not strong antioxidants, so they are normally used in combination with synergistic compounds. A recent patent by Chang and Wu (49) demonstrated that tocopherols with ascorbic acid, citric acid, and lecithin were effective as antioxidants in menhaden oil as well as vegetable oils. Many spices and herbs, such as sage, rosemary, and clove contain natural antioxidants. Herbalox, a natural seasoning from spices that have antioxidant activity is presently on the market.

### Antifoams

Two types of surface-active agents are used for foams in the food industry. A defoamer is used to decrease foam after it is formed, while an antifoam agent is used to prevent or reduce the foam from being formed. Dimethylpolysiloxane is the most widely used antifoam agent in fats and oils.

### Light

Previously, it was mentioned that light can initiate photo-oxidation, especially when photosensitizers such as chlorophyll are present. Compounds that absorb light and reduce its oxidation effect are known as quenchers. Several quenchers that have been shown to be effective in SBO are β-carotene and tocopherol. In addition to quenchers, light can be prevented from reaching the product by using dark bottles as in the beer industry, or light-proof packages for potato chips.

## Conclusion

As in any paper presenting data and facts, some important aspects may have been overlooked. Nevertheless, I have attempted to show that oil quality is influenced by many factors. It is possible that the quality of life also may be influenced by many of these same factors. The fundamental causes of human and animal aging have been accorded a great amount of descriptive research, but little knowledge of the fundamental causes of aging has come to light.

In recent years, some research has attempted to determine whether the uncontrolled presence of free radicals in the animal body has an effect on the organism's aging. From results to date, reactions of free radicals have been implicated in the mechanism of aging at the molecular level. Recent interest in natural antioxidants by consumers showed they are just as concerned about their bodies as they are about the quality of the foods they consume.

We may not be able to stop aging but hopefully, we can slow down the process. Although it is not the intent of this chapter to address aging of living organisms, some of the factors that affect oil quality and stability may also prove important in affecting survival curves. Feeding studies done with vitamin E, 2-mercaptoethylamine HCl, BHT, and EQ have already shown positive effects against some specific phenomena associated with the aging process and/or age-inducing stress.

## References

1. Watkins, L.R., Johnson, W.H., and Doty, S.C. "Extrusion- Expansion of Oilseeds for Enhancement of Extraction, Energy Reduction and Improved Oil Quality." in *Proceedings of the World Congress on Vegetable Protein Utilization in Human Foods and Animal Feedstuffs,* edited by Thomas H. Applewhite, The American Oil Chemists' Society, Champaign, IL, 1988, pp. 41–46.
2. Anderson, G. "Oilseeds Extraction and Meal Processing." in *Proceedings of the World Congress on Vegetable Protein Utilization in Human Foods and Animal Feedstuffs,* edited by Thomas H. Applewhite, The American Oil Chemists' Society, 1988, pp. 47–52.
3. Lusas, E.W., and Watkins, L.R. 1988. *J. Am. Oil Chem. Soc.* 65:1109.
4. Singleton, J.A., and Pattee, H.E. 1981. *J. Am. Oil Chem. Soc.* 58:873.
5. Jamieson, G.S., and McKenney, R.S. 1935. *Oil & Soap* 12:70.
6. Chu, Y.-H., and Lin, J.Y. 1993. *J. Am. Oil Chem. Soc.* 70:1263.
7. Lai, M.T., Lin, W.M., Chu, Y.-H., Chen, S.L.Y., and Kong, K.S. 1989. *J. Am. Oil Chem. Soc.* 66:565.
8. Swift, C.E., Mann, G.E., and Fisher, G.S. 1944. *Oil & Soap* 21:317.
9. Strecker, L.R., Maza, A., and Winnie, G.F. "Corn Oil-Composition Processing and Utilization," in *World Conference Proceedings of Edible Fats and Oils Processing: Basic Principles and Modern Practices,* edited by D.R. Erickson. The American Oil Chemists' Society, Champaign, IL, 1989, pp. 309–323.
10. Al-Kahtani, H.A.M. 1984. *J. Am. Oil Chem. Soc.* 61:94.
11. Evans, C.D., Schwab, A.W., Cooney, P.M., and Cowan, J.C. 1954. *Abstract* 84, Minnesota AOCS Meeting.
12. List, G.R., Mounts, T.L., Warner, K., and Heakin, A.J. 1978. *J. Am. Oil Chem. Soc.* 55:277.
13. List, G.R., Mounts, T.L., and Heakin, A.J. 1978. *J. Am. Oil Chem. Soc.* 55:280.
14. Beal, R.E., Lancaster, E.B., and Brekke, O.L. 1956. *J. Am. Oil Chem. Soc.* 33:619.
15. Segers, J.C. 1983. *J. Am. Oil Chem. Soc.* 60:262.
16. Mistry, B.B., and Min, D.B. 1987. *J. Food Sci.* 52:831.
17. Min, D.B., and Jung, M.Y. In Chapter 14 of *Flavor Chemistry of Lipid Foods,* edited by D.B. Min and T.H. Smouse. The American Oil Chemists' Society, Champaign, IL, 1989.
18. Canessa, C.E., Patterson, R.E., Berg, K., and Seybold, J.C. The 84th AOCS Annual Meeting and Exposition, April 25–29, 1993. *Abstract II* 1.
19. Toeneboehn, G.J. *Environmental Contributions of Silica Refining,* presented at The Fourth Latin American Meeting on Fats & Oils, November 22–26, 1992, Rosario, Argentina.
20. Goss, W.H. *The German Oilseed Industry,* Hobart Publishing Co., Washington, DC, 1947.

21. Patterson, H.B.W. *Bleaching and Purifying Fats and Oils: Theory and Practice,* The American Oil Chemists' Society, Champaign, IL, 1992.
22. Cowan, J.C. 1966. *J. Am. Oil Chem. Soc.* 43:300A.
23. Taufel, K., Franzke, U., and Heder, G. 1959. *Fette Seif Ans.* 61:1225.
24. Usuki, R.; Suzuki, T.; Endo, Y. and Kaneda, T. 1984. *J. Am. Oil Chem. Soc.* 61:785.
25. Gundstone, F.D. 1984. *J. Am. Oil Chem. Soc.* 61:441.
26. Mag, T.K. "Bleaching—Theory and Practice," in *Edible Fats and Oils Processing: Basic Principles and Modern Practices,* edited by D. Erickson, The American Oil Chemists' Society, Champaign, IL, 1989, p. 108.
27. Rost, H.E. 1963. *Fette Seifen. Ans.* 65:463.
28. Endo, Y., Usuki, R., and Kaneda, T. 1984. *J. Am. Oil Chem. Soc.* 61:781.
29. Welsh, W.A., Bogdanor, J.M., and Toeneboehn, G.J. "Silica Refining of Oils and Fats," in *Edible Fats and Oils Processing: Basic Principles and Modern Practices* edited by D. Erickson, The American Oil Chemists' Society, Champaign, IL, 1989.
30. List, G.R., and Erickson, D.R. "Storage, Handling and Stabilization" in *Handbook of Soy Oil Processing and Utilization,* edited by D.R. Erickson, E.H. Pryde, O.L. Brekke, T.L. Mounts, and R.A. Falb. The American Soybean Association and The American Oil Chemists' Society, Champaign, IL, 1980.
31. Klagge, P., and Sen Gupta, A.K. 1990. *Fat Sci. Technol.* 92:315.
32. Scavone, T.A., and Braun, J.L. U.S. Patent 4,789,554, Dec. 6, 1988.
33. Chang, S.S., and Kummerow, F.A. 1953. *J. Am. Oil Chem. Soc.* 30:251.
34. Johnson, O.C., Chang, S.S. and Kummerow, F.A. 1953. *J. Am. Oil Chem. Soc.* 30:317.
35. Chang, S.S., and Kummerow, F.A. 1953. *J. Am. Oil Chem. Soc.* 30:403.
36. Chang, S.S., and Kummerow, F.A. 1954. *J. Am. Oil Chem. Soc.* 31:324.
37. Witting, L.A., Chang, S.S., and Kummerow, F.A. 1957. *J. Am. Oil Chem. Soc.* 34:470.
38. Holm, U., Ekbom, K., and Wade, G. 1957. *J. Am. Oil Chem. Soc.* 34:606.
39. Brekke, O.L., Mounts, T.L., and Pryde, E.H. In *Handbook of Soy Oil Processing and Utilization,* edited by D.R. Erickson, E.H. Pryde, O.L. Brekke, T.L. Mounts, and F.A. Falb. The American Soybean Association and the American Oil Chemists' Society, Champaign, IL, 1980, p. 557.
40. Bishov, S.J., Henick, A.S., Giffee, J.W., Nil, I.T., Prell, P.A., and Wolf, M. 1971. *J. Food Sci.* 36:532.
41. Hu, K.H., and Breyer, J.D. 1970. *Food Technol.* 24:705.
42. Mickel, H.S., Horbar, J., Gewirtz, M., and Ansari, R. 1976. *La Riv. Ital. Delle Sost. Grasse* 53:312.
43. Robinson, H.E., and Black, H.C. 1945. *Ind. Eng. Chem.* 37:217.
44. Golumbic, C., Schepartz, A.I., and Daubert, B.F. 1946. *Oil & Soap* 23:380.
45. Smith, R.L. *The Sequestration of Metals,* Chapman & Hall, Ltd., 1959.
46. Furia, T.E., "Sequestrants in Foods," in *Handbook of Food Additives,* 2nd Edition, The Chemical Rubber Co., CRC Press, Boca Raton, FL, 1975.
47. Furia, T.E., "Sequestrants," in *Encyclopedia of Food Science,* edited by M.S. Peterson and A.H. Johnson, Avi, Westport, CT, 1978.
48. Houlihan, C.M., and Ho, C.T., "Natural Antioxidants" in *Flavor Chemistry of Fats and Oils,* edited by D.B. Min and T.H. Smouse, The American Oil Chemists' Society, Champaign, IL, 1985.
49. Chang, S.S., and Wu, K. U.S. Patent 5,077,069, Dec. 31, 1991.

Chapter 3

# Organization of a Sensory Evaluation Program

**L.J. Malcolmson**

Department of Foods and Nutrition, University of Manitoba, Winnipeg MB R3T 2N2, Canada.

## Introduction

Sensory evaluation is a multidisciplinary science that uses humans to measure the acceptability and sensory properties of food and other materials. Sensory characteristics important in food products include attributes of appearance, taste, odor, and texture. The use of humans as measuring devices is necessary because only humans can define what is "ideal" or "acceptable." Because there is no instrument that can replicate or replace the human response, sensory evaluation is a necessary and vital component in any quality assessment program.

The success of any sensory evaluation program begins by having a plan, being organized, and ensuring that the testing process follows accepted procedures and practices (1). This includes consideration of the experimental design, sensory test method, selection of panelists, testing facilities, sampling and serving procedures, and data analysis.

### *Design of a Sensory Program*

The organization of a sensory evaluation program must have full management support to ensure the success of the program. Without management approval, sensory professionals will be limited in their ability to conduct sensory tests, undermining the credibility of the program. Thus, the first step in organizing sensory activities begins with a plan to develop management's awareness of the value of sensory information (2). Stone and Sidel (1) offer a comprehensive discussion of how this can be achieved.

### *Professional Staff*

In addition to recognizing that a sensory program is needed in the organization, management must employ an individual who is capable of managing the sensory program if development of a sensory department is to be successful. Since sensory evaluation is the application of knowledge and skills from several scientific disciplines, including food science, psychology, physiology, mathematics, and statistics, it may be necessary to upgrade the skills of an employee or to create a new position and recruit an individual with the requisite training. The individual must be given full responsibility to manage the program and must determine how sensory information could be used within the organization. It is also important that this individual be given the time to manage and operate the sensory program. This cannot be done if the individual has other responsibilities within the organization.

## Sensory Methods

In the past, all sensory evaluation methods have been classified incorrectly as subjective techniques. Sensory tests can be divided into two general categories: affective (or subjective) and analytical (or objective). Affective tests measure an individual's opinion about a product, usually by having panelists rate their preference for a product or the amount they like or dislike a product. In contrast, analytical or objective tests measure the presence and/or intensity of specific sensory attributes in a product. These tests are done under controlled testing conditions, utilizing trained panelists to minimize the effects that psychological and physical conditions can have on panelists' judgments. According to Larmond (3), when these procedures are followed, analytical sensory tests meet the criteria to be considered objective methods. Watts et al. (4) have reclassified sensory tests as either consumer-orientated or product-oriented tests, better reflecting the purpose of the two major classes of sensory testing methods.

The sensory method chosen for a test is dependent on the type of information required. The appropriate method can only be selected after the objectives of the test are clearly defined. As illustrated in Table 3.1, there are a substantial number of test methods available, and it is the responsibility of the sensory professional to be familiar with them in order to apply them properly. Familiarity with the test methods also reduces the likelihood of reliance on a single method (1). Lawless and Claassen (5) provide a decision-making tree or flowchart for selecting the appropriate sensory method, based on test objectives and research questions. A more thorough discussion concerning test methodology is available elsewhere (1,6–12).

## Factors Influencing Sensory Measurements

Standard procedures have been developed for conducting sensory panels in order to minimize psychological errors and physical testing conditions that can influence human judgment. By controlling these factors while planning and conducting sensory experiments, reliable results can be achieved.

### *Expectation Error*

Expectation error occurs when panelists are given information about the samples prior to the test. This information can influence results, since panelists usually find what they expect to find. To control for this error, only give panelists enough infor-

**TABLE 3.1**
Summary of Product-Oriented and Consumer-Oriented Procedures

| | | |
|---|---|---|
| **Product-Oriented Tests** | | |
| Discrimination | | |
| | Threshold | |
| | | Method of limits |
| | | Average error |
| | | Frequency methods |
| | Difference | |
| | | Paired comparison |
| | | Duo-trio |
| | | Dual-standard |
| | | Triangle |
| Quantitative | | |
| | Scaling | |
| | | Ranking |
| | | Category |
| | | Magnitude estimation |
| | Duration | |
| | | Time-intensity |
| Qualitative | | |
| | Descriptive analysis | |
| | | Flavor profile |
| | | Texture profile |
| | | Quantitative descriptive analysis |
| | | Hybrid methods |
| | | Deviation from reference |
| **Consumer-Oriented Tests** | | |
| Acceptance | | |
| | Accept/not accept | |
| Preference | | |
| | Forced choice selection | |
| Hedonic | | |
| | Degree of like/dislike | |

mation for them to perform the task. Code samples so that panelists cannot identify them. Any persons who are involved with the test should not be included on the panel. Keep the testing area free from noise and distraction and separate from the food preparation area.

### *Suggestion Effect*

The reaction of other panel members can influence the response of a panelist. To minimize this error, separate panelists using individual booths. Do not permit discussion during testing.

### *Stimulus Error*

Stimulus error occurs when panelists are influenced by irrelevant sample differences, such as differences in volume, size, and color of the samples. To minimize this error, the samples must be prepared, sampled, and served under identical conditions. Unwanted color differences can be masked by using colored lights or filters.

### *Positional Bias or Order Effect*

The order in which samples are evaluated can influence a panelist's judgment. The first sample of a series of samples is often scored higher than expected regardless of the product. To minimize this effect, serve samples to panelists in randomized orders to ensure that samples are presented in different positions. By randomizing the sample order, positional bias can be balanced for the panel as a whole.

### *Central Tendency Error*

This error is characterized by panelists scoring products in the midrange of a scale, that is, a panelist avoids the extreme ends of the scale. To minimize this error, again present samples in randomized orders, since this effect is more noticeable for the first sample. Familiarize panelists with the products, test methods, and the testing procedures during training.

### *Contrast and Convergence Errors*

Contrast error can occur when two samples are markedly different from each other. Panelists may exaggerate the difference in their scores. Convergence is the opposite effect, brought about by samples masking smaller differences between other samples. This causes scores to converge. To minimize contrast and convergence effects, randomize the order of sample presentation for each panelist.

## Sensory Testing Facilities

An important requirement of a sensory evaluation program is the availability of a facility specifically designed for sensory testing. This is necessary to ensure that tests are run efficiently and conducted under controlled conditions with minimal distractions. Without proper controls, reliable results may not be achieved. Permanent testing facilities provide the best testing environment. If this is not feasible, it is possible to adapt existing facilities for sensory testing. The types of tests to be conducted, the amount of testing to be done, and the space and resources available will be deciding factors in the design of the sensory laboratory (4). Basic requirements for any facility include preparation area, booth area, and training area. A more thorough discussion regarding the design of sensory testing facilities including illustrations of possible layouts is available elsewhere (1,13).

### Preparation Area

The area allocated for food preparation is dependent on the type and number of products to be tested. Generally, a well-equipped kitchen is often adequate, provided there is sufficient counter space for tray layout. It is important that the area provide refrigerated and frozen storage to store or hold samples. If sample heating or cooking is required, installation of electric or gas ranges and microwave ovens will be necessary. For cleaning purposes, sinks and dishwashers are needed. Installation of distilled or filtered water should be considered for panelists rinse water, preparation of solutions, cooking, and rinsing dishes if tap water imparts an odor or flavor. Storage cupboards for supplies should be constructed above and below the serving counters. Ventilation hoods with exhaust fans should be installed to remove odors from the preparation area and to prevent these odors from spreading to the booth area. Also, air conditioning should be installed to maintain a constant temperature.

### Booth Area

*Permanent Booths.* The aim in designing the testing area is to provide the panelist with the optimal setting for unbiased judgment (3). To achieve this goal, individual booths are constructed so that panelists can assess samples without influence from other panelists. The booth area should be separate from the preparation area. The usual arrangement is the construction of a panel room adjacent to the preparation area. This permits the samples to be passed through the wall of the preparation area to the panelists. Six to eight booths should be adequate to run most sensory programs. Each booth should be equipped with a counter, stool or chair, a pass-through opening to the preparation area, individual lighting, and an electrical outlet for warming trays or water baths. Booths may also be equipped with individual sinks for expectoration, but this is not recommended due to sanitation, noise, and odor problems. The use of covered disposable containers for expectoration is a more acceptable alternative (1). Some booths are computerized so that panelists utilize a computer screen, keyboard, and mouse rather than a paper ballot to record their judgments.

A booth width of between 67 and 80 cm (27–32") and a counter depth of between 45 and 60 cm (18–24") is recommended (13). Counter height should be same as the counter on the other side of the pass-through wall to permit ease of passing sample trays from one side to the other. Standard counter height is 90 cm (36"). Partitions between booths should be at least 90 cm (36") high (4) and should extend 45 cm (18") beyond the edge of the countertop to provide panelist privacy (13). Stone and Sidel (1) recommend that a soffit be constructed 120 cm (48") above the booth counter that is about 45 cm (18") wide. This is used to accommodate the booth's lighting fixtures and ventilation ducts. Adequate space must be provided in the booth area to allow panelists to enter or leave without disturbing other panelists during evaluations. Pass-throughs from the food preparation area should be large enough to allow samples and trays to be passed easily to the panelist. An opening 40 cm (16") wide and 30 cm (12") high should be adequate (4). It should also be flush with the counter. The opening can be fitted with a sliding, bread-box, hinged, or flip-up door. However, bread-box, hinged, and flip-up openings will require more counter space.

Each booth should have its own lighting fixture to ensure uniform light distribution. Incandescent or fluorescent lighting can be selected. A combination of the two offers the greatest testing flexibility, but cost and space requirements may not permit this choice. Incandescent lights can be dimmed to offer a range of light intensities but generate heat, a consideration when designing ventilation and air-conditioning systems. Fluorescent lights do not generate heat, are less costly to operate, and are recommended where color evaluation is critical, since bulbs which more closely match natural daylight are available. If masking color differences among samples is desired, colored filters, colored incandescent bulbs, or sodium lights can be used. Oil samples are often evaluated under red lights to mask color differences.

The panel room should be maintained at a comfortable temperature. The room should be air conditioned, and in some cases it may be necessary to control humidity. Adequate ventilation for the panel room is the most critical design aspect. A slight positive air pressure should be maintained in the booth area to prevent infiltration of external odors. Individual exhaust ventilation ducts should be located in each booth. Air turnover in the booths should occur at least every 30 sec. ([illegible]). Recirculated and makeup air should pass through activated carbon filters to remove odors.

Walls and booths should be constructed of opaque, nonreflecting material, which is neutral in color. Odor-free materials that are easily cleaned should be used for floors and counter tops.

*Temporary Booths.* If an area specifically designed for sensory testing is not available, a temporary area can be set up, provided that distractions, noise, and odors are kept to a minimum. A conference room or a lunch/coffee room not in use at the time might be suitable if odors are eliminated. Portable, odor-free partitions made of wood, metal, or heavy cardboard can be used to provide individual workstations for panelists.

### Training Area

For product-orientated testing it is necessary to have an area where panelists can meet with the sensory professional for training and discussion. The room must be separate from the food preparation area, so that noise and cooking odors do not interfere with the panelists' task. Ideally, this room should be adjacent to the preparation and booth areas. A comfortable well-lit room with good ventilation and a large table for 6–12 panelists is required. Additional tables near electrical outlets may be needed to hold extra samples. A chalkboard, whiteboard, or flip chart should be provided to facilitate discussion. A bulletin board should also be available for posting results.

## Samples

### *Sample Preparation and Presentation*

Samples should be prepared and presented to panelists under identical conditions. It is recommended that at least 30 g of solid food or 15 mL of liquid foods be served (7). It may be necessary to serve some foods with a carrier or to use dilutions. For example, crackers may be used as a carrier for the evaluation of margarine. However, it is generally better to avoid the use of a carrier if possible, since they have flavor and textural characteristics of their own which can interfere with the panelists' evaluations of the intended product. Through training, panelists can usually overcome any aversion they might have to sampling a product without a carrier. Products such as spices, hot sauces, or crude oil usually require dilution prior to testing, but the effect of the dilution medium should be considered.

#### ***Serving Temperature***

All samples must be presented at the same temperature and generally should be served at the temperature at which the food is usually consumed. Soft margarines are served at 4-5°C whereas hard margarines can be served at either 4-5°C or 22-24°C. It is recommended that vegetable oils be evaluated at a temperature of 50°C (14) since this temperature brings out their characteristic odor or flavor. Individual electric warming trays equipped with a water bath are ideal for keeping oil samples warm. Alternatively, aluminum blocks can be heated to the desired temperature prior to serving samples for evaluation.

#### ***Serving Containers***

The size and shape of containers will vary with the products being tested. Glass beakers, bottles, or petri-plates can be used, as can disposable paper or plastic containers. It is imperative that the serving containers impart no odor or flavor of their own. Samples must be served in identical containers. Lids should be used to prevent contamination of food, and in the case of odor evaluation to permit volatiles to build in the headspace. The American Oil Chemists' Society (14) recommends oil for flavor evaluation be prepared by placing 10 mL of oil in a 50 mL glass beaker covered with a watch glass. In our laboratory, oil for odor evaluation is prepared by placing 10 g of oil into a 125 mL glass jar containing 30 g of glass beads. Jars are then capped with teflon-lined plastic screw caps (15).

### Sample Coding and Serving Orders

Samples should be coded in such a way that they impart no information to the panelist that might influence their judgment (expectation error). For this reason it is recommended that samples be coded with three-digit random numbers as opposed to single digit or letter codes. Beckman et al. (16) found that color coding did not bias results and may be considered as an alternative coding system. However, this method cannot be used with panelists who exhibit difficulties in color perception or in testing which is done under colored lights.

### Reference Samples

The use of reference samples or standards is critical in both the training of panelists and calibrating of panelists during testing. Reference samples are used during training to help panelists develop the terminology they need to describe the samples. They are also used to anchor endpoints and midpoints of attribute-rating scales. The presence of reference samples during the testing period permits calibration of panelists before evaluation of the coded samples, thereby ensuring that panelists score the coded samples more consistently. This is especially critical when sensory testing is conducted over a long period of time (weeks or months). Reference samples can be either an internal reference (selected from the samples to be tested) or an external reference (chemical or alternative food product). For example, an internal reference might be a fresh oil sample selected to anchor the bland endpoint of an oil intensity scale, whereas an external reference might be a bland vegetable oil spiked with an artificial butter flavoring to anchor the intense endpoint of a buttery intensity scale. Regardless of the product used, it is important that the selected reference clearly illustrates the attribute that it was meant to represent, and that it is stable and will not change in character over the course of the sensory test. For further discussion on reference samples the reader is referred to a paper by Rainey (17). Chapters 4 and 5 provide information about reference samples that can be used in the testing of oils and fat-containing foods.

### Number of Samples

The number of samples that can be evaluated in one session is dependent on the nature of the product and on the amount of training and experience of panelists. Sensory fatigue may result from too many samples. In our laboratory, panelists evaluate no more than four oil or margarine samples per test session. This is due to the nature of the product, and the number of reference samples the panelists are also required to evaluate.

### Rinsing

Room temperature water is generally provided for rinsing. Warm water (38–40°C) is necessary to clear the mouth if fats or oils are being evaluated. Unsalted soda crackers or white bread can also be used. In some cases it may be necessary to specify a waiting time between samples. Samples can be swallowed or expectorated, but panelists should decide on a consistent technique. Fats and oils are usually expectorated.

## Selection and Training of Panelists

### *Recruiting and Screening of Potential Panelists*

Panelists for trained panels can be recruited from the organization where the testing is being conducted. Provided there is management support, most employees are willing to participate if they feel their contribution is important. Potential panelists should be asked to complete a simple questionnaire to determine their likes and dislikes of foods, their level of interest in participating in panels, any food restrictions or allergies, and times available for panels. This information is helpful for the sensory professional to screen individuals in terms of motivation, availability, and health. Generally, candidates should be in good health and should not be prone to frequent colds. Certain medications may alter a panelist's sensitivity to taste and odor compounds. Age, sex, and smoking habits may be other factors to consider in selecting panelists, although they are not critical. High level of vocabulary development is desirable but not essential since this can be improved with training.

The next step is to screen panelists for normal sensory acuity. This can be done by having panelists complete several sensory tests designed to identify their ability to recognize basic tastes, common odors, and textural characteristics. More details of screening tests used in the evaluation of panelists are available elsewhere (18).

### *Selection of Panelists*

The final selection of panelists for a specific test should be based on their ability to detect differences in the food to be evaluated and to be consistent in their judgments. This is often determined by triangle tests (7). For example, in the selection of panelists to serve on oil panels in our laboratory, oil samples of known differences are presented to potential panelists. Two of the coded samples are the same, and one is different. Panelists are asked to identify which sample is different. We generally give three different sets of three samples, and then repeat the entire exercise on the following day. By providing three sets we gain a measure of the panelist's ability to discriminate a range of samples of known differences, and by repeating the exercise we gain a measure of the panelist's ability to reproduce judgments. A group of 20–25 people is generally screened in order to select a panel of 8–12 individuals suitable for training. It is recommended that 10–12 panelists be trained, since some panelists may be lost during the study due to illness or changes in job-related priorities. Panelists showing good discrimination for one type of product may not show good discrimination for another product. Therefore it is necessary to perform these tests prior to the start of any new sensory testing.

### *Training Panelists*

The performance of individual panelists, and the panel as a whole is improved through training. The goal of the training program is to help panelists make valid, reliable, and objective judgments free of personal preference or bias. A discussion of results directed by the sensory professional accompanies each training session. Through discussion, panelists learn an understanding of the task, the ballot/scoring system, and the terminology. Over time, panelists learn to develop standardized methods of evaluation, so that panelists' responses become consistent and agree with each other.

Training exercises should be designed to teach panelists about the attributes they will be evaluating. This is done by introducing reference samples that illustrate each attribute. Panelists then evaluate coded samples for these attributes. The coded samples used in these training exercises should include some of the samples they will actually evaluate in the test session and may include spiked samples prepared to demonstrate a variation of an attribute. Panelists should be exposed to the range of samples they are likely to encounter in the test sessions during the training program.

Depending on the organization, training sessions are designed to last 30–60 min/d over a period of 1–2 weeks. The exact length of time depends on the nature of the products to be tested, the number of attributes to be evaluated, and the past experience of the panelists. Training is completed when panelists are in agreement about the attributes to be evaluated and the placement of the references on the scale, are comfortable with the evaluation procedure, can discriminate among samples of known difference, are in agreement as to the ordering of these samples on the scale, and can repeat their judgments. Panelist input throughout the training period is vital to the sensory professional, since it provides information necessary for establishing the number of samples to be evaluated in a given test session, finalizing the ballot, and selecting appropriate reference samples. The final training session should be a trial run of a test session.

### *Monitoring Panelists' Performance*

The sensory professional should monitor panelist performance as a group and individually. To determine this, a set of samples of known difference must be evaluated by the panelists on several occasions. Data is analyzed using analysis of variance to determine significant variation among samples and among panelists. Lack of significance among samples of known difference indicates the need for further training, since panelists are not responding to differences that are known to exist. Significant differences among panelists are not unexpected, but the sensory professional must ensure that panelist differences are due to use of scale (panelists may use different portions of the scale to express their judgments) and not due to different ordering of samples on the scale. The American Society of Testing and Materials (18) can provide further guidance on panel monitoring.

### *Motivation of Panelists*

It is important to maintain the interest and motivation of panelists throughout training and test sessions, since panelists who are highly motivated perform better than non-motivated panelists. During training, panelists receive feedback about their performance during the discussion period, but this can also be done individually. Feedback is not generally given during the test sessions but should be provided at the conclusion of the study. Panelists should be rewarded for their participation and not for correct scores. Food or beverage rewards can be given at the conclusion of each training and test session; or a luncheon, party, or gift at the conclusion of all sessions. This latter approach is used in our laboratory, when we also provide information about the study including test objectives, products tested, and group results. It is important that panelists know that their contribution was appreciated, otherwise recruiting for future panels will be difficult.

## Experimental Design

The experimental design selected for a test will influence the accuracy of the results. The design chosen will depend on the objectives of the test and will dictate the serving order of the samples, as well as, the most appropriate analysis to satisfy the test objective (1). There are many types of experimental designs to choose from, including completely randomized, balanced complete block, and balanced incomplete block designs. Readers are advised to consult Chapter 11 or a statistical textbook or a statistician for selecting the simplest and most efficient design to meet the test objectives.

Key elements to good experimental design include randomization, blocking, and replication. Randomization is done to minimize the effects of uncontrolled sources of variation or error and to eliminate bias. The procedure involves ordering sample treatments such that each treatment has an equal chance of being selected. Items typically randomized in a sensory test include the order of sample presentation and the assignment of a sample to an oven for cooking.

Blocking is used to increase the power of an experiment by removing known sources of variability from the estimate of error. Blocks may include panelists, replications, storage interval, or anything that is a known source of variation. By grouping experimental units into blocks, a more accurate measure of pure or experimental error is obtained.

Replication involves repeating the experiment under identical conditions. It improves the reliability and validity of test results and is necessary to provide an estimate of experimental error. The number of replications is determined by time, cost, and sample constraints. The more replications completed, the better the estimate of experimental error and the more reliable the test. This book contains further information on statistical treatment of data (Chapter 11).

## References

1. Stone, H. and Sidel, J.L. *Sensory Evaluation Practices,* Academic Press Inc., San Diego, CA, 1993.
2. Eggert, J. 1989. *J. Sensory Studies* 3:161.
3. Larmond, E. "Sensory Evaluation Can Be Objective," in *Objective Methods in Food Quality Assessment,* edited by J.G. Kapsalis, CRC Press, Boca Raton, FL, 1987.
4. Watts, B.M.; Ylimaki, G.L.; Jeffery, L.E. and Elias, L.G. *Basic Sensory Methods for Food Evaluation,* International Development Research Centre, Ottawa, Canada, 1989.
5. Lawless, H.T. and Claassen, M.R. 1993. *Food Technol.* 47:139.
6. Amerine, M.A.; Pangborn, R.M. and Roessler, E.B. *Principles of Sensory Evaluation of Food,* Academic Press, New York, NY, 1965.
7. American Society for Testing and Materials, *STP 434 Manual on Sensory Testing Methods,* ASTM, Philadelphia, PA, 1968.
8. American Society for Testing and Materials, *MNL 13 Manual on Descriptive Analysis Testing For Sensory Evaluation,* ASTM, Philadelphia, PA, 1992.
9. Institute Food Technologists Sensory Evaluation Division. 1981. *Food Technol.* 35:50.
10. Meilgaard, M.; Civille, G.V. and Thomas, B.T. *Sensory Evaluation Techniques,* CRC Press Inc., Boca Raton, FL, 1991.
11. Moskowitz, H.R. *Applied Sensory Analysis of Foods,* Vol. I, II, CRC Press Inc., Boca Raton, FL, 1988.
12. Poste, L.M.; Mackie, D.A.; Butler, G. and Larmond, E. *Laboratory Methods for Sensory Analysis of Food,* Agriculture Canada, Ottawa, Canada, 1991.
13. American Society for Testing and Materials, *STP 913 Physical Requirement Guidelines for Sensory Evaluation Laboratories,* ASTM, Philadelphia, PA, 1986.
14. The American Oil Chemists' Society, Method Cg 2-83 in *Official and Tentative Methods of the American Oil Chemists' Society,* 4th ed., AOCS, Champaign, IL, 1989.
15. Przybylski, R.; Malcolmson, L.J.; Eskin, N.A.M.; Durance-Tod, S.; Mickle, J. and Carr, R. 1993. *Lebensm. Wiss. u. Technol.* 26:205.
16. Beckman, K.J.; Chambers IV, E. and Gnagi, M.M. 1984. *J. Food Sci.* 49:1115.
17. Rainey, B.A. 1986. *J. Sensory Studies* 1:149.
18. American Society for Testing and Materials, *STP 758 Guidelines for the Selection and Training of Sensory Panel Members,* ASTM, Philadelphia, PA, 1981.

Chapter 4

# Sensory Evaluation of Oils and Fat-Containing Foods

**Kathleen Warner**

Food Quality and Safety Research, USDA, ARS, NCAUR, 1815 N. University St., Peoria, IL 61604, USA.

## Introduction

Sensory evaluation is the ultimate method to assess oil quality and stability, because no instrumental or chemical methods have been developed that have taken the place of our noses, taste buds, and sensory receptors. Instrumental and chemical methods measure portions of the factors, such as decomposition products, that comprise quality, but no method other than sensory analysis integrates all factors into a total perception of flavor intensity or overall quality. We are able to correlate gas chromatographic volatiles, peroxide values, and other analyses with sensory data, but while these measurements do correlate, they are not exact substitutes for smell and taste. Unfortunately, sensory analysis is performed by humans who may not have the reproducibility of a gas chromatograph; however, trained, experienced sensory testers operating under controlled conditions can provide accurate, reliable data.

This chapter will show how sensory analysis has a place in quality control, product development, and research ranging from lab-bench facilities operating on a limited budget to state-of-the-art operations with large sensory staffs and computerized data handling. Any sensory operation, no matter how small, can provide valuable information depending on the objectives. This chapter contains basic information for using sensory evaluation to determine product quality and stability. Included are guidelines for getting a sensory operation started, as well as, upgrading an existing program. This information may also be important to people wanting to collaborate or contract with a sensory operation to conduct tests.

### *What Is Sensory Evaluation?*

Before discussing whether or not sensory evaluation is needed or how to use it, it must be defined. Sensory evaluation is a scientific discipline used to evoke, measure, analyze, and interpret reactions to food and material characteristics as they are perceived by senses of sight, smell, taste, touch, and hearing. Sensory evaluation of oils is limited mostly to the senses of taste and smell. Some of the other terminology and definitions for sensory science are presented in Appendix A. Definitions and reference standards for specific flavors and odors are given in Appendix B.

### *Is Sensory Evaluation Needed?*

You may be at the point of determining whether or not sensory analysis is necessary. A major question to ask is "Are our existing analyses telling us all there is to know about our samples/products?" There are no specific guidelines to follow; however, the following steps may help with decision making.

First, list any problems encountered in your processing operation or research when analysis of product quality and/or stability may be needed. Examples of common problems are presented in Table 4.1. Next list the methods available to address these problems. Typical instrumental, chemical, and sensory analyses are given in Table 4.2. Review the list of problems and the methods used to measure and determine if any of the situations are not adequately covered by instrumental and chemical analyses.

In an industrial setting, if a change in ingredients occurs and the results of the analyses (i.e., peroxide value, or hexanal content) are identical to the product before the change in ingredients, there may be no need for sensory analysis. On the other hand, if consumer complaints appear with products that instrumental and chemical analyses have given clearance for, then sensory analyses may be needed to pinpoint problems. Also, keep in mind that all indicators of off-flavors and -odors are not detectable by instrumental and chemical analyses. The best example of this is low levels of linolenic acid oxidation when oils develop a distinctive fishy flavor. Peroxides and volatiles are usually low in this case, but flavor quality is poor because of the fishy off-flavor. Hydrogenation flavor is not detected by standard oxidation measures, except sensory analysis.

**TABLE 4.1**
Potential Problems in Production or Research

| |
|---|
| New supply source |
| Inconsistent quality of supply |
| Process change |
| Ingredient change/new additive |
| Poor storage stability/shelf life |
| Off-flavors/odors in product |
| New variety |
| Alternative antioxidants |

**TABLE 4.2**
Analyses to Measure Quality and/or Stability

| |
|---|
| Sensory |
| Gas chromatographic volatiles |
| Peroxide values |
| Active oxygen method (AOM)/Rancimat |
| Free fatty acids |
| Color |
| Conjugated diene |
| Oxygen uptake |

Ideally, sensory evaluation should be incorporated as part of an oil- or fat-containing food analysis program that also includes other methods listed in Table 4.2. We have found that the first three tests listed, sensory evaluation, gas chromatographic volatiles, and peroxide value, provide the most information about the quality of salad oils (1). No test should stand alone as a measure of quality or stability. Finally, just because a test is fast does not mean that it is a good indicator of quality or stability.

If an in-house, trained, and experienced panel is not an option, then contracting or collaborating with an established panel may be an acceptable alternative. In a quality control situation, all personnel should be trained to recognize a good quality product and to detect off-quality samples. Even in a research setting, scientists should at least smell all samples to learn differences in odor quality, especially if tasting a sample is not practical or safe.

### Applications for Sensory Evaluation

In addition to measuring product quality and stability, sensory evaluation can be used for a wide variety of situations, such as those listed in Table 4.3.

### Benefits from Sensory Evaluation

In the early 1940s, researchers at the USDA Northern Regional Research Center were investigating the off-flavor problems of soybean oil. Since they were having trouble analyzing the oil by conventional instrumental or chemical methods, they established a taste panel for vegetable oils because the grassy and beany flavors in soybean oil could only be detected by humans. Later, when considerable research was directed at identifying and solving the flavor problem of soybean oil, a trained, experienced panel helped pinpoint such problems as trace metal contamination, light sensitivity, poor shelf life/storage stability, and improper processing (3–9). Later research showed that hydrogenating soybean oil increased its stability; however, the processing imparted a characteristic flavor to the oil that was only detectable by sensory panelists and could not be detected by peroxide or volatile compound analyses (10).

At this point, if it has been decided that sensory evaluation can play a significant role in your operation, then the following information on establishing a sensory evaluation program should be consulted. Information will be included later on upgrading an existing program.

**TABLE 4.3**
Sensory Evaluation Applications

| |
|---|
| Monitor product quality |
| Assess alternative processes |
| Assess alternative additives/formulations |
| Monitor stability |
| Evaluate packaging |
| Increase product knowledge |
| Develop product descriptions |

### *Types of Sensory Evaluations*

The typical analyses for odor and flavor of fats, oils, and fat-containing foods are listed in Table 4.4. Salad oils can be evaluated for odor or flavor as liquid oils after deodorization as finished products. Smelling or tasting an oil in liquid form at 50°C without dilution or combination with any other foods is the most sensitive sensory analysis. Previous research at our laboratory has shown that crude and partially processed oils can also be tasted, but must be diluted to make the sample palatable to panelists (11,12). Flavor characteristics of oil which were present in the crude, degummed, refined, and bleached stages were reliably detected by diluting the crude or partially processed oils in good quality oil at dilution levels of 1–5%. We were also able to use the flavor intensity and flavor characteristics of crude soybean oil to predict the storage stability of those same oils after they were deodorized and aged (12). Tasting bleached oils diluted to a 5:95 ratio with good quality oil also will provide excellent reference standards for characteristic off-flavors and odors in oils, such as the typical pine flavor detected in oxidized sunflower oil.

Oil-containing foods can also be evaluated for flavor and storage stability by a sensory panel. Methods were developed to prepare salad dressings and to evaluate the flavor deterioration of fresh and aged products (13). These sensory procedures can also be used for mayonnaise and pourable dressings. The flavor of fried food is an important sensory analysis both for foods that are consumed immediately after frying, such as French fried potatoes, and for those that are eaten after storage, such as potato chips. In the category of freshly fried food, the most commonly tested products are potatoes, chicken, and fish. Potatoes have an advantage over animal products because they do not contribute any fat to the frying system. For laboratories wanting to use a realistic frying schedule, rotation frying of potatoes, chicken, and fish is appropriate. To test the effect of storage on a fried product, a snack-type food, such as potato chips, can be prepared. In previous research, we have reported the use

**TABLE 4.4**
Typical Sensory Evaluations for Edible Oils and Fat-Containing Foods

| Flavor/Odor Analyses | Room Odor Analyses |
|---|---|
| Salad oils | Heated oil odors |
| Deodorized | Frying food odors |
| Partially processed | |
| (diluted in RBD oil) | |
| Fat-containing foods | |
| Salad dressing | |
| Mayonnaise | |
| Fried food | |
| French fried potatoes | |
| Chicken | |
| Fish | |
| Potato chips | |
| Bread cubes | |

of a model food, 1" squares of white bread, that can be fried and tested as a snack food, either immediately after frying or after aging (10).

Other researchers have also used bread squares effectively to test the frying stability of oils (14). The odors of heated fats and oils and frying fats and oils can be evaluated by a sensory panel. In the 1970s, Evans and co-workers (15,16) reported a method to evaluate the room odor of heated oils and the room odor of oils used for frying. The method, perfected through subsequent research, has been used to evaluate liquid frying oils, solid fats, and oils from modified seeds, such as high-oleic sunflower and low-linolenic acid soybean (10,17–20).

## Organization and Structure of Sensory Panels

### *Types of Sensory Panels*

The characteristics of the various sensory panels are presented in Table 4.5. The two general types are consumer and analytical. Consumer panels are important in market research to directly measure preference or acceptability of a food product by those buying food products. Consumers would probably not evaluate oils directly, but they can judge oil-containing foods, such as potato chips or salad dressings.

Trained, experienced subjects—an analytical panel—are needed to evaluate oils directly for their subtle flavors and distinctive off-flavors. The two basic types of analytical panels are difference and descriptive. A difference panel conducts simple evaluations by answering a question, such as "Is there a difference in flavor between these samples?" The level of training and experience is usually minimal; therefore,

**TABLE 4.5**
Types and Characteristics of Sensory Panels

| Types | Characteristics |
|---|---|
| Analytical | Trained testers<br>Normal sensory acuity<br>5–20 members |
| Difference<br>Triangle,<br>Duo-trio, ranking,<br>Paired comparison | Measures differences between samples |
| Descriptive<br>Scalar scoring,<br>Descriptive analysis | Measures intensity or quality |
| Consumer | Measures acceptance, preference, or like-dislike response<br>50+ untrained testers |

this type of panel is an excellent choice for those just wanting to start an oil panel. On the other hand, more information is usually needed about an oil than just knowing if it is different from another sample. A descriptive panel takes the next step in oil evaluation by reporting not only if there are differences between oils, but also what those differences are. Panelists are trained to rate the oils for quality or intensity based on precise scales, with reference standards that leave as little room as possible for subjective analysis. Panelists can also be taught to describe the odors and flavors that they detect and to rate the intensity levels of those odors/flavors.

### *Sensory Panel Requirements*

There are several levels of sensory panel development from a simple operation to a state-of-the-art facility. The main difference in the panel levels depends first on the purpose or focus of the sensory panel operation and, of course, the time and resources that the company or laboratory is willing to contribute. Table 4.6 presents four levels of sensory panel development with the requirements for each level.

### *The Basic Sensory Panel*

Any operation can establish a simple sensory panel to evaluate oils for quality control. The basic requirements include an individual with scientific training to direct the panel and to apply scientific methodology to evaluating the oils; an odor-free, quiet room; and a group of 5–12 people screened to taste oils. A training program in basic smelling and tasting techniques, followed by instruction on simple difference

**TABLE 4.6**
Sensory Panel Requirements

| Factors | Levels | | | |
|---|---|---|---|---|
| | Basic | Upgrade 1 | Upgrade 2 | Upgrade 3 |
| Focus: | Quality Control | Effect of processing changes<br>Storage stability | Product development<br>Correlate instrumental and chemical analysis | Research on cause-and-effect |
| Personnel: | Scientist<br>+ consultant | Food technologist<br>+ technician<br>+ consultant | Sensory scientist<br>+ technicians | Sensory scientist<br>+ technicians |
| Subjects: | Screened<br>Trained in:<br>• Mechanics of smell and taste<br>• Difference testing | Also trained in:<br>• Scalar scoring | Also trained in:<br>• Descriptive analysis | Also trained in:<br>• Descriptive analysis<br>Experienced |
| Recording method: | Paper Ballot | Paper Ballot | Paper Ballot | Computer I/O |

tests will start the panel. Although this panel will not solve major problems, such as determining cause and effect relationships between volatiles and flavor, they will be able to determine whether or not a new supply of oil for food processing is different from the previous supply, for example. As sensory panel operations are upgraded, panelists can be trained for scalar scoring and descriptive analysis. These techniques will not only help determine differences between oils but also provide information as to the cause of the variations. In the case of a company that uses vegetable oil in food processing and must screen bulk-oil shipments, a bench-top operation can be established in which a 500 mL sample of oil is taken to evaluate flavor and to determine peroxide value as outlined in the following procedure.

### *Procedure*

Follow AOCS method Cg 2-83 (2) of heating oils to 50°C. Trained laboratory personnel taste the oil and compare it with a control, using the triangle test form (Appendix D). The control and the new sample are not identified. The control, which should be kept frozen until use, should meet the company's specifications for a good quality oil. The data from the triangle test is statistically analyzed by consulting tables in sensory manuals (22,23). In most cases, at least two-thirds of the panel should detect a difference before the difference is considered statistically significant. As the taste tests are done, the peroxide values can be determined. The process should take only 15 min. after the oils are heated.

*Upgrade 1.* The focus of the sensory panel at this level includes monitoring the effect of processing changes or ingredient changes and evaluating storage stability. The objective at this level goes beyond simply determining whether or not there are differences to identifying the amount and type of difference. This panel should be trained to rate samples on scoring scales—intensity or quality. Examples of intensity- and quality-rating scales are in Appendix D.

The panel leader should have a background in food science or food technology, and college courses in sensory analysis or on-the-job training in sensory evaluation methodology. A sensory evaluation consultant can also be utilized to provide answers. If the volume warrants, a technician can be hired to help with the preparation and presentation of samples to the panel and cleanup.

*Upgrade 2.* Upgrading the sensory panel operations to the next level is usually necessary if specific work, such as product development, is conducted or if the research includes correlating sensory with instrumental and chemical analyses of oils. The panel should receive training in descriptive analysis to be able to fully describe the odors and flavors detected in the products. A sensory scientist who has a degree in sensory evaluation or who has experience leading panels is invaluable in directing the operations at this level. Qualifications of a sensory scientist include a knowledge of psychology, physiology, statistics, and chemistry. The facilities should also be upgraded to include an area set aside for the panel room with adjoining area for the preparation of the oils. Information on the specific needs of a taste-panel room are given in Chapter 3, as well as in sensory evaluation handbooks (22,23) and the AOCS Recommended Practice Cg 2-83 (2). The use of techniques to mask color differences, for example by red-colored fluorescent bulbs, are effective in hiding color differences of almost all oils and oil-containing foods, unless of course, there is a large difference in color intensity.

*Upgrade 3.* Laboratories needing this level of sensory panel operation evaluate a wide variety of oils and oil-containing foods that require several groups of trained panelists. To help with the work load, a manager of sensory operations may be needed, as well as, the capacity to record panel data in computer storage media. Panelists can use various methods to record scores and descriptions, such as a touch-sensitive computer screen, keyboard, or mouse. A printout of the data with statistical calculations provides immediate feedback to the panel leader. The screen/keyboard/mouse can be operated through a mainframe computer or by a small personal computer. A separate computer screen that monitors the panelists' scores as they are entered into the computer can be installed in a room next to the panel area. This allows the panel leader to view the results and permits the panelists to check their own completed evaluations with those of other testers. An example of a printout used at our research center and explanations of computer programs to calculate, store, and retrieve data has been published (24).

*Risk Reduction.* Studies involving human subjects may need review by internal or external committees for assurance that the subjects are not considered to be at risk. Even if no such reviews exist, the panel leader should take precautions to ensure the safety of judges by knowing the treatments the samples have undergone and the presence of any possible risks from the samples. Judges should sign voluntary consent forms that contain the following information: project title, treatments tested, risks and benefits of the testing, and precautions to be taken. Information on the treatments may be proprietary; therefore, judges do not have to be given complete details, only those that may cause risks such as any non-GRAS (generally recognized as safe) additives. Examples of precautions include instructing judges not to swallow samples and to rinse their mouths thoroughly after tasting. Judges could evaluate only odor, rather than flavor if there is any question of sample safety.

## Oil-Panel Evaluation Methods

### *Liquid Oils*

Salad-type oils that are liquid at room temperature can be evaluated for odor and flavor to determine their initial quality and/or flavor stability after storage. The AOCS (2) has a recommended practice (Cg 2-83) for the preparation and presentation of oils for sensory testing, as does the American Society of Testing and Materials (ASTM [21]). These practices include basic information, such as serving containers, oil temperature, and heating procedures for the oils. The AOCS recommended practice also includes two scoring scales based on a 10-point system for rating the quality or intensity of liquid vegetable oils (Appendix D). The intensity scale is based on a 10 = bland, 1 = strong scale. This is recommended for rating oils that have blandness as their ultimate goal. Oils, such as soy, sunflower, cottonseed, canola (LEAR), and safflower, that have little or no flavor immediately after proper processing and deodorization are rated on this scale. On the other hand, a quality scale (ratings of excellent to bad) should be used for oils, such as corn, peanut, and olive, that have naturally distinctive, although desirable flavors. If a company has a variety of bland and flavorful oils, then the quality scale may be a realistic choice rather than using the two scales.

Reference standards for the intensity scale should include examples of oils with weak, moderate, and strong overall flavor intensity. A standard for a completely bland oil may be difficult to find; therefore, the panelists should have a mental standard for blandness—no flavor. Reference standards for the quality scale should include examples of what the company defines as excellent, good, fair, poor, and bad quality. A list of reference standards is presented in Appendix B.

Reference standards for intensity and quality levels may be verified by instrumental and chemical analyses. For example, an oil with a peroxide value of 0–0.2 is usually good quality or has weak overall flavor intensity. A volatile compound profile by gas chromatographic analysis may also be used to check standards. In using these scales, the panelists should be instructed to evaluate the overall intensity or overall flavor quality of the sample first. Then they should evaluate the oil for individual flavors and their intensity levels. In actual practice, experienced panelists will go through this process simultaneously rather than separating the two steps. The score sheet or computer-scoring screen should be set up for the panelists to record their overall scores first followed by the ratings for the individual descriptions.

### Descriptions

Typical oil descriptions range from no flavor (bland) to strong flavors (rancid and painty [Table 4.7]). Panelists should not only be instructed to note any odors/flavors detected in the samples, but also to indicate the intensity level of each description. At our laboratory, we ask panelists to rate the odors/flavors of salad oils on a scale of 0 = none, 1 = weak, 2 = moderate, 3 = strong. A weighted average of the intensity levels of each description is then calculated.

$$\sum \frac{(\text{Rating Value} \times \text{Number of Responses})}{(\text{Number of Panelists})}$$

Quantifying the description-intensity levels provides valuable information that might not be readily seen from the overall scores. This data can also be used in statistical calculations, such as correlation coefficients with levels of volatile compounds. Descriptions can also be rated on scales of other sizes, such as on the 10-point scale that we use for salad dressing or room odor (Appendix D). The range of the scale chosen depends on the ability of the panelists and the need for determining finer differences between samples. The 3-point scale simplifies the process of assigning intensity levels to descriptions.

Panel leaders and panelists should not only know the characteristic odors/flavors of each oil type but also something about the source of these odors/flavors (Table 4.7). Soybean oil evaluated immediately after sufficient processing and deodorization has little or no odor/flavor and is usually described as bland or as slightly nutty or buttery. If the processing or deodorization is not adequate, then the soybean oil may have a weak grassy or beany flavor.

As the oil oxidizes, odors/flavors become stronger. In the early stages of oxidation, soybean oil may have weak to moderate buttery odor/flavor. As oxidation continues, odors/flavors develop that are typical of a crude or partially processed oil. Sometimes this is referred to as a reverted flavor or retrograded flavor (27). For soybean oils, these odors/flavors are grassy and beany. Flavors detected in later stages of oxidation of soybean and canola oil include both rancid and painty, because both linoleic and linolenic acid are present in these oils.

Oils, such as palm or peanut, that contain no linolenic acid, do not develop a painty flavor since this is characteristic of linolenate-containing oils, such as soybean and canola. Some inexperienced oil panelists can confuse a strong rancid odor/flavor with painty. Oils, such as cottonseed, with high levels of linoleic acid will develop a predominant rancid flavor in more highly oxidized samples. Frankel (28,29) pre-

**TABLE 4.7**
Characteristic Flavor Descriptions of Vegetable Oils[a]

| Processing Step and Storage | Vegetable Oil Type | | |
|---|---|---|---|
| | Soybean | Canola[b] | Sunflower |
| Crude or refined, bleached oil (diluted 5:95 in deodorized oil) | Hay, grassy, green, grassy, beany | Cabbage, sulfur, grassy/green | Pine/cedar, weedy, acrid |
| Deodorized | | | |
| Fresh/unaged | Nutty, buttery | Nutty, buttery | Nutty, buttery |
| Storage in dark, slight oxidation | (Flavors are similar to crude or bleached oils) | | |
| moderate to high oxidation | Rancid, painty | Rancid, painty, fishy | Rancid |
| Storage in light | "light struck" (grassy, sour, metallic, buttery) | Buttery, grassy, metallic | Stale, sour |

[a]All deodorized oils contain 100 ppm citric acid.
[b]Low-erucic acid rapeseed oil (LEAR).

sented reviews of fat and oil oxidation which explained the mechanism of formation and the flavor significance of oxidation products.

### Room Odor/Frying Oils and Fats

Methods to evaluate the odor characteristics of heated and frying oils were developed in the 1970s by Evans and co-workers (15,16). The initial method included simply heating a pan of oil in an empty laboratory and having panelists enter the room and rate the odor in the room for overall quality and individual odor characteristics. This method was upgraded by constructing small rooms with controlled air flow and temperature (17,25). Panelists were instructed to rate the overall intensity of room odor and the intensities of the individual odors detected.

The intensity-scoring scale (Appendix D) is a 10-point scale (no odor to strong odor). Oils are heated to 190°C, and the weakest overall intensity odor for a sample is usually 3.0 for oils such as high-oleic sunflower or well-processed low-linolenic acid soybean oil. Fried food is the predominant odor of these oils. Canola oil would have a strong overall odor intensity with predominant odors of fishy, acrid, and burnt. Hydrogenated oils have a characteristic "hydrogenation" odor that we train our panelists to recognize and to describe the individual odors that make up this complex odor. Odors/flavors such as fruity, flowery, and waxy are typical of hydrogenated oils. Appendix B contains standards for odors of heated oils. Details on the design of a room-odor facility as well as sensory analysis procedures for heated oil evaluations are available (25,26).

### Oil-Containing Foods

The sensory evaluation of oil-containing foods is an important analysis because of limited instrumental and chemical methods to test these foods. Methods to evaluate starch-based salad dressings (13) are also appropriate for mayonnaise. Pourable salad dressings can be either evaluated directly or tasted in a model system such as the emulsion method proposed by Stone and Hammond (30).

Although the emulsion method was suggested as an alternative to direct tasting of oils to prevent taste fatigue, it could also be an effective method to test pourable salad dressing. In our research, we have found that testing starch-based salad dressings without a carrier is acceptable to panelists. Some laboratories may want to use carriers with as little flavor as possible, such as unsalted crackers or lettuce. During panel training for these evaluations, panelists should be taught to recognize the flavor characteristics of a good quality product. They should also learn to rate the intensities of the complex variety of tastes and flavors present, such as sweet, sour, stale, or rancid.

The products should be evaluated with a quality-scoring scale since a good quality salad dressing is not bland. An example of a score sheet for salad-dressing analysis is in Appendix D. A good quality control should be given to the panelists at the beginning of each testing session. Examples of off-flavored products can be created by preparing the dressings with oils of different quality or by aging dressings in a shelf-life test.

Fried foods can be evaluated either immediately after frying (i.e., french fries, chicken, and fish) or after storage, such as snack foods (i.e., potato chips). The panelists evaluate the flavor of the food and determine whether or not there are any flavors or off-flavors from the oil or fat. Since blandness is not a desired characteristic of fried foods, a quality scale is appropriate (Appendix D). A good quality control should be given to the panelists at the beginning of each testing session as a reference. We generally use a product fried in cottonseed oil as an example of a good quality fried food. As an alternative to frying a typical food product, we used 1" bread squares for panel evaluation immediately after frying and after storage (10,20).

## Data Interpretation

### *Intensity Scale for Salad Oil*

Interpretations of the intensity-scoring scale for salad oils are presented in Table 4.8. Sensory scales should be linear. The intensity scale was designed to measure flavor changes as the oil oxidizes. With increasing oxidation, flavor intensity not only increases, but the type of flavors detected progressively changes. Most freshly deodorized oils have little or no flavor, but as the oil oxidizes, the flavors change or are masked by oxidatively derived flavors. For example, as soy oil oxidizes, flavors progressively change from nutty and buttery to grassy, hay, beany, and finally to rancid and painty.

The intensity scale is applicable for all oils that have no distinctive natural flavors immediately after deodorization. Freshly deodorized oils, such as soy, sunflower, canola, and cottonseed, should have little or no flavor or odor if the original crude oil quality was satisfactory and if all processing was conducted properly. Panelists evaluating a fresh, laboratory-processed soybean oil on an intensity scale

**TABLE 4.8**
Interpretation of Oil Intensity Scale[a]

| Score | Intensity Level | Characteristic |
|---|---|---|
| 10 | Bland | No flavor detected |
| 9 | Trace | Detectable, but too weak to identify |
| 8 | Faint | Typical of most freshly deodorized oils |
| 7 | Slight | Typical of most commercial oils |
| 6 | Mild | Typical of oils with PV < 5 |
| 5 | Moderate | |
| 4 | Definite | Typical of oils with PV > 5 |
| 3 | Strong | |
| 2 | Very strong | Typical of oils with PV > 10 |
| 1 | Extreme | |

[a]Intensity scale measures intensity of overall flavor.

would usually rate the oil a 10 for odor, since no detectable odor would be present. The flavor score would usually be a 9, since a trace of flavor is detectable but is too weak to be identified. This difference between the odor and flavor scores exists because panelists are usually able to detect sensory characteristics more easily by the tasting techniques than by just smelling the oil.

A commercially processed soybean oil evaluated soon after processing usually receives a score of 8 on an intensity scale, indicating faint but detectable flavor or flavors, such as nutty or buttery. After this oil has been bottled and stored on a grocery shelf for a few months, the flavor score of the oil is 7, indicating a low level of oxidation. Typical flavors may be buttery, grassy, or beany. An intensity score of 7 for a fresh or slightly aged oil is usually considered the cutoff point for a good oil.

Oils that are scored a 6 usually have peroxide values between 1 and 5, although off-flavors from processing, such as hydrogenation, may cause the intensity of the flavor to increase. A score of 6 is considered the minimum score for acceptability after aging, as in the Schaal oven test (4 d at 60°C) or after 4 months at ambient temperature (25°C). Oils that are scored 5 or below are usually oxidized, as is indicated by peroxide values above 5. The score would continue to decrease as the oxidation process proceeded.

It is important to point out that oils with an overall intensity level of 7 or 8 would not be described as weakly rancid or painty because these flavors are typical of later stages of oxidation and therefore, generally do not exist by themselves, but rather in conjunction with other off-flavors, such as grassy. Flavors such as grassy or beany precede the development of rancid flavor, so by the time an oil reaches the peroxide level necessary for a rancid or painty flavor, other flavors are present. On the other hand, strong nutty or buttery flavors are generally not present in oils, because these flavors either are masked by other off-flavors or the compounds responsible for the flavors are decomposed in oxidized oils.

### *Quality Scale for Salad Oils*

The interpretation of the quality scale given in Appendix D is in Table 4.9. This scale is appropriate for oils with distinctive natural flavors, such as corn, peanut, and olive oils. The quality characteristics used for this scale should be set by individual laboratories according to the type of oil. For example, a well-processed corn oil usually has a weak corny flavor, no off-flavors and, therefore, can be considered good quality. Individual standards can be changed if a company considers an oil with a moderate corn flavor as indicative of good quality. The quality level of the corn oil would be rated down if the oil contained any off-flavors from processing, such as a weak to moderate burnt flavor. The quality would also decrease as the oil underwent oxidative deterioration and either lost some or all of its characteristic natural flavor or developed flavors indicative of oxidation. The score would decrease further depending on the intensity of the off-flavors. To remind the panel of the characteristics of a good quality product, a standard control sample can be presented during each evaluation.

**TABLE 4.9**
Interpretation of Oil Quality Scale[a]

| Score | Quality Level | Characteristic |
|---|---|---|
| 10 | Excellent | |
| 9 | Good | |
| 8 | | Typical of good quality oil; weak characteristic flavors, no off-flavors |
| 7 | Fair | Weak off-flavor or some loss of characteristic flavor |
| 6 | | |
| 5 | Poor | Moderate off-flavors |
| 4 | | |
| 3 | Very poor | Strong off-flavors |
| 2 | | |
| 1 | Bad | |

[a]Quality scale measures quality of overall flavor.

Some scales used by the oil industry are a combination of the intensity and quality scale. Panelists are instructed to taste the oil, determine the flavors in the oil, then assign a score based on the type and intensity of the oil flavor. For example, an oil with a weak melon flavor is rated a 5 and an oil with a weak painty flavor is rated a 4.

This type of scale was developed because in actual practice this is the progression of oil deterioration; however, it is not an appropriate scale for two reasons. First, a scale based on descriptions is not linear. An intensity scale or a quality scale based on physical examples of specific intensity levels or of specific quality levels has equal intervals between numbers. According to Bodyfelt et al. (34), a linear scale is a prerequisite for statistical analysis. In the example given previously, the difference in score between an oil with a weak melon flavor and a weak painty flavor is probably more than one unit; therefore, this combined scale based on descriptions is not linear, and any statistical results would have little meaning. The second deficiency of this scale is that it depends on determining the type of oil flavor in order to assign a flavor score. The accuracy of this scale depends on the ability of all panelists to correctly and consistently identify flavors detected, which is the most difficult task for an oil panelist. A scale based on descriptions is not recommended for oil evaluations, but the use of either an intensity scale or a quality scale with reference standards for intensity or quality levels is recommended. The quality and intensity scales in Appendix D were used successfully in an AOCS collaborative study in 1988 (AOCS, unpublished data).

### *Intensity Scale for Room Odor*

The characteristics of the room-odor intensity levels are in Table 4.10. A room with no oil odor is rated as 0, and the score is increased with increasing odor intensity. Scores of 1–3 are indicative of weak overall odors. Cottonseed oil heated at 150°C is typical of an oil with weak overall room-odor intensity, usually scored as 3.0. Scores

**TABLE 4.10**
Interpretation of Room-Odor Intensity Scale[a]

| Score | Intensity Level | Characteristic |
|---|---|---|
| 0 | None | No odor detected |
| 1 | | |
| 2 | Weak | |
| 3 | | Typical of cottonseed oil heated at 150°C |
| 4 | | Typical of high oleic (80%) oils heated at 190°C |
| 5 | Moderate | Typical of cottonseed oil heated at 190°C |
| 6 | | Typical of hydrogenated (IV = 90) soybean oil |
| 7 | Strong | Typical of soybean or rapeseed oils heated at 190°C |
| 8 | | Typical of soybean or rapeseed oils heated at 240°C |
| 9 | | |
| 10 | | |

[a]Room-odor intensity scale measures intensity of overall room odor and individual odors.

of 4–6 are indicative of moderate overall intensity. Corn oil heated at 190°C is typical of an oil with a moderate overall intensity. Most oils are rated in this area of the scale. Oils with strong odor intensity would be rated as 7–10. The strongest odor standard that we use is a canola oil heated to 240°C, which receives an average score of 8.

## *Quality Scale for Oil-Containing Foods*

This quality scale (Table 4.11) is similar to the quality scale for oils. The characteristics for each type of food need to be determined by each laboratory/company. For example, a potato chip that is considered a good quality product with a characteris-

**TABLE 4.11**
Interpretation of Quality Scale For Oil-Containing Foods[a]

| Score | Quality Level | Characteristic |
|---|---|---|
| 10 | Excellent | |
| 9 | Good | |
| 8 | | Typical of good quality product; weak characteristic flavors, no off-flavors |
| 7 | Fair | Weak off-flavor or some loss of characteristic flavor |
| 6 | | No characteristic flavor; weak off-flavors |
| 5 | Poor | Moderate off-flavors |
| 4 | | |
| 3 | Very Poor | Strong off-flavors |
| 2 | | |
| 1 | Bad | |

[a]Quality scale measures quality of overall flavor.

tic potato-chip flavor and no off-flavors would be rated as an 8. As the potato chips age and deteriorate during storage, a loss of typical potato-chip flavor is noted, and the flavor quality score decreases. As off-flavors begin to develop, the score continues to decrease with increasing amounts of off-flavors. In fried foods, one of the first off-flavors to appear is a stale flavor. The subsequent off-flavors are usually the same as in oxidized oils.

### *Sources of Information*

The following is a list of publications from various organizations with general information about operating sensory programs as well as specific information about oil panels.

American Oil Chemists' Society, Champaign, Illinois, USA.

Recommended Practice Cg 2-83; *Flavor Chemistry of Fats and Oils; Handbook for Soy Oil Processing.*

American Society for Testing and Materials, Philadelphia, PA, USA.

*Standard Practice for the Bulk Sampling, Handling, and Preparing Edible Vegetable Oils for Sensory Evaluation* E-1346-90; *Basic Principles of Sensory Evaluation,* STP 433, 1968; *Correlating Sensory Objective Measurements, New Methods for Answering Old Problems,* STP 594, 1976; *Consumer Sensory Evaluation,* STP 682, 1979; *Manual on Sensory Testing Methods,* STP 434, 1968; *Guidelines for the Selection and Training of Sensory Panel Members,* STP 758, 1981; *Physical Requirement Guidelines for Sensory Evaluation Laboratories,* STP 913, 1986; *Manual on Descriptive Analysis Testing for Sensory Evaluation,* MNL 13, 1992.

### References

1. Warner, K., Frankel, E.N., and Mounts, T.L. 1989. *J. Am. Oil Chem. Soc.* 66:558.
2. *Official and Tentative Methods of the American Oil Chemists' Society,* 4th ed., American Oil Chemists' Society, Champaign, IL, 1989, Methods Cd 8-53, Cg 1-83, and Cg 2-83.
3. Dutton, H.J., Moser, H.A., and Cowan, J.C. 1947. *J. Am. Oil Chem. Soc.* 24:21.
4. Dutton, H.J., Schwab, A.W., Moser, H.A., and Cowan, J.C. 1948. *J. Am. Oil Chem. Soc.* 25:385.
5. Evans, C.D., Cooney, P.M., Moser, H.A., Hawley, J.E., and Melvin, E.H. 1952. *J. Am. Oil Chem. Soc.* 29:61.
6. Dutton, H.J., Evans, C.D., and Cowan, J.C. 1953. *Trans. Am. Assoc. Cer. Chem.* 11:116.
7. Evans, C.D., Frankel, E.N., Cooney, P.M., and Moser, H.A. 1960. *J. Am. Oil Chem. Soc.* 3724:452.
8. Moser, H.A., Evans, C.D., Cowan, J.C., and Kwolek, W.F. 1965. *J. Am. Oil Chem. Soc.* 42:30.
9. Cowan, J.C. 1965. *Food Tech.* 19:107.
10. Frankel, E.N., Warner, K., and Moulton, K.J. 1985. *J. Am. Oil Chem. Soc.* 62:1354.

11. Frankel, E.N., Warner, K., and Klein, B.P. 1988. *J. Am. Oil Chem. Soc.* 65:147.
12. Warner, K., Frankel, E.N., and Moulton, K.J. 1988. *J. Am. Oil Chem. Soc.* 65:386.
13. Warner, K., Frankel, E.N., Snyder, J.M., and Porter, W.L. 1986. *J. Food Sci.* 51:703.
14. Miller, L.A., and White, P.J. 1988. *J. Am. Oil Chem. Soc.* 65:1324.
15. Evans, C.D., Moser, H.A., List, G.R., Dutton, H.J., and Cowan, J.C. 1971. *J. Am. Oil Chem. Soc.* 48:711.
16. Evans, C.D., Warner, K., List, G.R., and Cowan, J.C. 1972. *J. Am. Oil Chem. Soc.* 49:578.
17. Mounts, T.L. 1979. *J. Am. Oil Chem. Soc.* 56:659.
18. Warner, K., Mounts, T.L., and Kwolek, W.F. 1985. *J. Am. Oil Chem. Soc.* 62:1483.
19. Mounts, T.L., Warner, K., List, G.R., Kleiman, R., Fehr, W.R., Hammond, E.G., and Wilcox, J.R. 1988. *J. Am. Oil Chem. Soc.* 65:624.
20. Warner, K. 1988. *J. Am. Oil Chem. Soc.* 65:520.
21. American Society for Testing and Materials, *Standard Practice for the Bulk Sampling, Handling, and Preparing of Edible Oils for Sensory Evaluation,* Philadelphia, PA, 1990.
22. American Society for Testing and Materials, STP 434, *Manual on Sensory Testing Methods,* Philadelphia, PA, 1968.
23. Canada Department of Agriculture, *Methods for Sensory Evaluation of Food,* Ottawa, Canada, 1978.
24. Warner, K., Ernst, J.O., Boundy, B.K., and Evans, C.D. 1974. *Food Tech.* 28:42.
25. Mounts, T.L., and Warner, K. "Evaluation of Finished Oil Quality," in *Handbook of Soy Oil Processing and Utilization,* edited by D.R. Erickson, E.H. Pryde, O.L. Brekke, T.L. Mounts, and R.A. Falb, American Soybean Association and American Oil Chemists' Society, Champaign, IL, 1980.
26. Warner, K. "Sensory Evaluation of Flavor Quality of Oils" in *Flavor Chemistry of Fats and Oils,* edited by D. Min and T.H. Smouse, American Oil Chemists' Society, Champaign, IL, 1985.
27. Smouse, T.H. "Flavor Reversion of Soybean Oil" in *Flavor Chemistry of Fats and Oils,* edited by D. Min and T.H. Smouse, American Oil Chemists' Society, Champaign, IL, 1985.
28. Frankel, E.N. "Chemistry of Autoxidation: Mechanism, Products and Flavor Significance" in *Flavor Chemistry of Fats and Oils,* edited by D. Min and T.H. Smouse, American Oil Chemists' Society, Champaign, IL, 1985.
29. Frankel, E.N. 1982. *Prog. Lipid Res.* 22:1.
30. Stone, R., and Hammond, E.G. 1983. *J. Am. Oil Chem. Soc.* 60:1277.
31. American Society for Testing and Materials, STP 758, *Guidelines for the Selection and Training of Sensory Panel Members,* Philadelphia, PA, 1981.
32. Voirol, E., and Daget, N. 1986. *Lebensm.-Wiss. u.-Technol.* 19:316.
33. Berglund, B., Berglund, U., and Lindvall, T. 1976. *Psychol. Rev.* 83:432.
34. Bodyfelt, F.W., Tobias, J., and Trout, G.M. *Sensory Evaluation of Dairy Products,* Van Nostrand Reinhold, New York, NY, 1988.
35. Warner, K., Evans, C.D., List, G.R., Boundy, B.K., and Kwolek, W.F. 1974. *J. Food Sci.* 39:761.

# Appendix A

## *Definitions*

**Analytical panel:** Group of trained, experienced panelists who rate products by discriminatory and descriptive sensory methods. A group of 12–15 people is recommended. Panelists are trained to measure intensity or quality as objectively as possible with no value judgments.

**Consumer panel:** Group of untrained individuals who rate products for preference or on a like-dislike scale.

**Flavor:** Integrated attributes of taste and odor. Odor is detected by smelling the volatiles of the sample, whereas the flavor is detected after placing the sample in the mouth. Volatiles go to the olfactory bulb either nasally (through the nose) or retronasally (through the back of the oral cavity). In most cases, flavor is a more sensitive analysis of sensory characteristics, because all volatiles are directed toward the olfactory nerves and are not lost between the headspace of the sample container and the nose. Panelists almost always report a stronger flavor from a sample in their mouth than they do from a sample that is only smelled.

**Intensity scale:** Method to rate the intensity or strength of an odor, taste, or flavor on a scale, usually 0–10.

**Objective:** Unbiased report of stimulus from instrument or from trained, experienced sensory panelists using discriminatory or descriptive tests.

**Odor:** Sensation from stimulus of olfactory receptors through the nose by volatile compounds. Thousands of volatile compounds produce simple and complex odors detected in food products.

**Organoleptic:** Archaic term meaning to sense with nose and tongue. This term was used before scientists knew that the sensory mechanisms of perception are located at higher levels of the nervous system than the level of tongue and nose. The word sensory is more accurate.

**Quality scale:** Method to rate the esthetic properties of a product, usually on a 1–10 scale of bad to excellent.

**Sensory evaluation:** Scientific discipline used to evoke, measure, analyze, and interpret reactions to the characteristics of foods and materials as they are perceived by senses of sight, smell, taste, touch, and hearing. Sensory evaluation of oils is limited mostly to the senses of taste and smell.

**Subjective:** Biased report from an individual on preference or like-dislike scales. Consumer panels produce subjective analyses of flavor.

**Taste:** Sensation from stimulus of taste buds on tongue, there are four basic tastes—sweet, salt, sour, and bitter. Taste is not usually applicable to oils, because oils do not have these tastes with the possible exception of bitter.

## Appendix B

### *Lexicon for Fats, Oils and Oil-Containing Foods with Definitions (Def) and Standards (Std)*

**Acrid:**
Def: Aromatic characteristic of burnt oil, pungent, sharp.
Std: Canola oil heated to 240°C, acrolein.
**Beany:**
Def: Aromatic characteristic of soybeans.
Std: Cooked soybeans, not raw or green.
**Bland:**
Def: *NO* aromatics or tastes perceptible, an oil should not be described as bland with a slight nutty flavor.
**Burnt:**
Def: Aromatic characteristic of overheating/cooking.
Std: Crude corn oil (odor only).
**Buttery:**
Def: Aromatic characteristic of fresh, unsalted butter.
Std: 1% unsalted fresh butter/99% fresh oil free of off-flavors.
**Cardboard:**
Def: Aromatic characteristic of slightly oxidized oils or oil-containing foods.
Std: Water with cardboard or paper soaked in it for 30 min and filtered.
**Corny:**
Def: Aromatic characteristic of steeped corn and corn oil.
Std: Crude corn oil diluted in good quality deodorized corn oil at a 5:95 ratio.
**Fishy:**
Def: Aromatic characteristic of oxidized linolenate-containing oils, such as linseed, canola, or soy.
Std: Cod liver oil or canola oil heated to 190°C for 30 min.
**Fried Food:**
Def: Aromatic characteristic of food fried in hot oil.
Std: Deep fat fried potatoes (french fried potatoes).
**Fruity:**
Def: Aromatic characteristic of ripe noncitrus fruits.
Std: Olive oil.
**Grassy:**
Def: Aromatic characteristic of freshly mown green grass.
Std: Fresh cut green grass, hexanal.
**Green:**
Def: Aromatic characteristic of raw, unripe vegetables.
Std: Raw soybeans.

**Hay:**
Def: Aromatic characteristic of dried grass or hay.
Std: Crude soy oil from heat-processed beans (dilute 5:95 in fresh oil).
**Hully:**
Def: Aromatic characteristic of the coatings of grains.
Std: Peanut or sunflower hulls.
**Hydrogenated:**
Def: A combination of aromatics typical of hydrogenated oils (paraffin, fruity, flowery).
Std: Hydrogenated soy oil, with an iodine value < 100.
**Melon:**
Def: Aromatic characteristic of ripe watermelon or cantaloupe.
Std: 0.002 ppm *cis*-6-nonenal in fresh oil free of off-flavor.
**Metallic:**
Def: Aromatic characteristic of metal coins.
Std: 0.01% ferrous sulfate in distilled water.
**Musty:**
Def: Aromatic characteristic of damp basement.
Std: Geosmin (odor only).
**Nutty:**
Def: Aromatic characteristic of fresh, sweet, nuts.
Std: Freshly processed peanut oil.
**Overheated:**
Def: Aromatic characteristic of oil overheated during processing.
Std: Crude corn oil (odor).
**Oxidized:**
Def: General term to include aromatics characteristic of deteriorated or oxidized oils from slight to intense (e.g., buttery to painty). Not recommended as a specific description for oils or fat-containing foods.
**Painty:**
Def: Aromatic characteristic of linseed oil or of oxidized linolenate-containing oil (rapeseed/canola, soy) with a peroxide value greater than 10.0, not noted in non-linolenate oils, such as peanut or palm.
Std: Linseed oil.
**Pine:**
Ref: Aromatic characteristic of pine needles.
Std: Undeodorized sunflower oil in 5:95 ratio with fresh sunflower oil free of off-flavors.
**Rancid:**
Def: Aromatics characteristic of oxidized linoleate-containing oils.
Std: Oxidized safflower or sunflower oil with peroxide value greater than 5.0.
**Raw:**
Def: Aromatics characteristic of unprocessed grain or vegetables.
Std: Whole wheat flour, raw potato.

**Reverted:**

Def: General term denoting oil characteristic of an oxidation process known as reversion. Each oil type has specific flavors. Not recommended as a descriptor.

**Rubbery:**

Def: Aromatics characteristic of old rubber.

Std: Rubber stoppers for chemical glassware.

**Smoky:**

Def: Aromatics characteristic of overheated oil.

Std: Oil heated to 240°C.

**Sour:**

Def: Basic taste characteristic of acids, such as citric acid.

Std: 0.08% citric acid in water.

**Stale:**

Def: General term characteristic of old, unfresh product, with loss of original flavors, usually in initial stages of oxidation.

Std: Potato chips aged 2 weeks at 25°C.

**Sulfur:**

Def: Aromatics characteristic of sulfur compounds, noted in canola/rapeseed oil.

Std: Undeodorized canola/rapeseed oil diluted to 5 wt% in fresh oil.

**Waxy:**

Def: Aromatics characteristic of melted paraffin.

Std: Paraffin oil.

**Weedy:**

Def: Aromatics characteristic of green, nongrass weeds.

Std: Fresh cut green weeds.

**Woody:**

Def: Aromatics characteristic of dry wood.

Std: Sawdust or wood chips, popsicle sticks, peanut oil heated to 190°C for 30 min.

# Appendix C

Oil Attributes Characteristic of Unprocessed Oil (U), Freshly Processed Oil (F), Deteriorated Oil (D), or Origin Unknown (X)

| | Oil Type | | | | | | |
|---|---|---|---|---|---|---|---|
| Flavor Notes | Corn | Cottonseed | Peanut | Canola/Rapeseed | Safflower | Soy | Sunflower |
| Beany | | | | | | D/U | |
| Burnt | U | | X | | | U | X |
| Buttery | F | F | F | F | F | F | F |
| Cardboard | D | D | D | D | D | D | D |
| Corny | F | | | | | | |
| Fishy | | | | D | | D | |
| Grassy | X | | X | D | X | U/D | U/D |
| Hay | X | | X | | X | U/D | U/D |
| Hully | | | | | | | D |
| Nutty | F | F | F | | | F | |
| Painty | | | | D | | D | |
| Pine | | | | | | | U/D |
| Rancid | D | D | D | D | D | D | D |
| Raw | X | | X | X | | X | |
| Rubbery | X | | | X | | X | |
| Sour | | | X | X | | | |
| Sulfur | | | | U/D | | | |
| Waxy | X | X | | | X | | X |
| Weedy | X | | X | X | X | X | X |
| Woody | | | X | | X | X | X |

U = Characteristic of unprocessed or partially processed oil.
F = Characteristic of freshly processed, good quality oil.
D = Characteristic of deteriorated (from fair to bad quality) oil (no frying or high-temperature thermal abuse).
X = Unidentified origin.

Other flavors (pumpkin, melon, watermelon, petroleum, metallic, musty, etc.) may be present from several sources, including contamination and processing conditions.

# Appendix D

## *Score Sheets*

### Triangle Difference Test

Name: ____________  Date: ___________

**Instructions:** Taste the three samples in the order presented. Two of the samples are identical; one is different. Place a check mark by number of the different sample.

| Code Number | |
|---|---|
| 391 | ___________ |
| 376 | ___________ |
| 382 | ___________ |

### Fried Food Evaluation

Name: ________________  Date: ___________________

**Instructions:** Taste the samples in the order presented, beginning with the control (C). Rate the samples for individual flavors and their intensities and for overall quality. The control is an example of a good quality product (score = 8) with a moderate fried food flavor and no off-flavors.

Individual Flavor Intensity

| 0 | 1 | 2 | 3 | 4 | 5 | 6 | 7 | 8 | 9 | 10 |
|---|---|---|---|---|---|---|---|---|---|---|
| None | | Weak | | | Moderate | | | Strong | | |

| Sample Code | Intensity Scores | | | | | | |
|---|---|---|---|---|---|---|---|
| | Fried Food | Stale | Fishy | Hydrogenated | Waxy | Rancid | Painty |
| Control | ___ | ___ | ___ | ___ | ___ | ___ | ___ |
| 532 | ___ | ___ | ___ | ___ | ___ | ___ | ___ |
| 579 | ___ | ___ | ___ | ___ | ___ | ___ | ___ |

Overall Flavor Quality

| 1 | 2 | 3 | 4 | 5 | 6 | 7 | 8 | 9 | 10 |
|---|---|---|---|---|---|---|---|---|---|
| Bad | | Poor | | Fair | | Good | | Excellent | |

| Sample Code | Score |
|---|---|
| Control | _____ |
| 532 | _____ |
| 579 | _____ |

## AOCS Flavor Intensity Evaluation

**Directions:** Take 5–10 mL of warm oil into the mouth; pull air through the oil and exhale through the nose.
Overall Intensity Scores: Rate samples on a 10–1 scale.
Descriptions: Identify flavors and rate as weak (W), moderate (M), or strong (S).

| Intensity | | Overall Intensity Scores 492 | 716 | 258 | 931 |
|---|---|---|---|---|---|
| 10 | Bland | ____ | ____ | ____ | ____ |
| 9 | Trace | ____ | ____ | ____ | ____ |
| 8 | Faint | ____ | ____ | ____ | ____ |
| 7 | Slight | ____ | ____ | ____ | ____ |
| 6 | Mild | ____ | ____ | ____ | ____ |
| 5 | Moderate | ____ | ____ | ____ | ____ |
| 4 | Definite | ____ | ____ | ____ | ____ |
| 3 | Strong | ____ | ____ | ____ | ____ |
| 2 | Very strong | ____ | ____ | ____ | ____ |
| 1 | Extreme | ____ | ____ | ____ | ____ |

| Descriptions | Description Intensity | | | |
|---|---|---|---|---|
| Nutty | ____ | ____ | ____ | ____ |
| Buttery | ____ | ____ | ____ | ____ |
| Corny | ____ | ____ | ____ | ____ |
| Beany | ____ | ____ | ____ | ____ |
| Hydrogenated | ____ | ____ | ____ | ____ |
| Burnt | ____ | ____ | ____ | ____ |
| Weedy | ____ | ____ | ____ | ____ |
| Grassy | ____ | ____ | ____ | ____ |
| Rubbery | ____ | ____ | ____ | ____ |
| Melon | ____ | ____ | ____ | ____ |
| Rancid | ____ | ____ | ____ | ____ |
| Painty | ____ | ____ | ____ | ____ |
| Fishy | ____ | ____ | ____ | ____ |
| Other ____________ | ____ | ____ | ____ | ____ |
| Other ____________ | ____ | ____ | ____ | ____ |

## AOCS Flavor Quality Evaluation

**Directions:** Take 5–10 mL of warm oil into mouth; pull air through the oil and exhale through the nose. Rate samples for overall flavor quality on a 10-point scale; identify flavors and rate as weak (W), moderate (M), or strong (S).

| Quality | | Overall Quality Scores | | | |
|---|---|---|---|---|---|
| | | 492 | 716 | 258 | 931 |
| 10 | Excellent | ______ | ______ | ______ | ______ |
| 9 | Good | ______ | ______ | ______ | ______ |
| 8 | | ______ | ______ | ______ | ______ |
| 7 | Fair | ______ | ______ | ______ | ______ |
| 6 | | ______ | ______ | ______ | ______ |
| 5 | Poor | ______ | ______ | ______ | ______ |
| 4 | | ______ | ______ | ______ | ______ |
| 3 | Very Poor | ______ | ______ | ______ | ______ |
| 2 | | ______ | ______ | ______ | ______ |
| 1 | Bad | ______ | ______ | ______ | ______ |

| Descriptions | | Description Intensity | | | |
|---|---|---|---|---|---|
| Nutty | | ______ | ______ | ______ | ______ |
| Buttery | | ______ | ______ | ______ | ______ |
| Corny | | ______ | ______ | ______ | ______ |
| Beany | | ______ | ______ | ______ | ______ |
| Hydrogenated | | ______ | ______ | ______ | ______ |
| Burnt | | ______ | ______ | ______ | ______ |
| Weedy | | ______ | ______ | ______ | ______ |
| Grassy | | ______ | ______ | ______ | ______ |
| Rubbery | | ______ | ______ | ______ | ______ |
| Melon | | ______ | ______ | ______ | ______ |
| Rancid | | ______ | ______ | ______ | ______ |
| Painty | | ______ | ______ | ______ | ______ |
| Fishy | | ______ | ______ | ______ | ______ |
| Other | ____________________ | ______ | ______ | ______ | ______ |
| Other | ____________________ | ______ | ______ | ______ | ______ |

## Salad Dressing Evaluation

Name: ______________________ Date: ______________

**Instructions:** Taste each sample in the order presented, starting with the control (C).

Mark the intensity rating by putting the number of the sample along the po nt in the line for each flavor detected.

Rate the overall quality of the samples on the scale at the bottom of the sheet.

Intensity

Sweetness

Weak [ ] Strong

Sourness

Weak [ ] Strong

Grassiness

Weak [ ] Strong

Rancidity

Weak [ ] Strong

Paintiness

Weak [ ] Strong

Hydrogenated

Weak [ ] Strong

Staleness

Weak [ ] Strong

Other

Weak [ ] Strong

Overall quality

[ ]

Bad Poor Fair Good Excellent

## Room Odor Evaluation

Order No. ____________ Date ____________ Name ____________________

Directions: Circle number to indicate intensity level.

| | | | | | | | | | | | |
|---|---|---|---|---|---|---|---|---|---|---|---|
| Overall intensity | 0 | 1 | 2 | 3 | 4 | 5 | 6 | 7 | 8 | 9 | 10 |
| | | | Weak | | | Moderate | | | Strong | | |
| Fried food | 0 | 1 | 2 | 3 | 4 | 5 | 6 | 7 | 8 | 9 | 10 |
| Fishy | 0 | 1 | 2 | 3 | 4 | 5 | 6 | 7 | 8 | 9 | 10 |
| Hydrogenated | 0 | 1 | 2 | 3 | 4 | 5 | 6 | 7 | 8 | 9 | 10 |
| Burnt | 0 | 1 | 2 | 3 | 4 | 5 | 6 | 7 | 8 | 9 | 10 |
| Doughy | 0 | 1 | 2 | 3 | 4 | 5 | 6 | 7 | 8 | 9 | 10 |
| Acrid/pungent | 0 | 1 | 2 | 3 | 4 | 5 | 6 | 7 | 8 | 9 | 10 |
| Waxy | 0 | 1 | 2 | 3 | 4 | 5 | 6 | 7 | 8 | 9 | 10 |
| Other ________ | 0 | 1 | 2 | 3 | 4 | 5 | 6 | 7 | 8 | 9 | 10 |
| | | | Weak | | | Moderate | | | Strong | | |

Chapter 5

# Sensory Evaluation of Margarine

**M. Vaisey-Genser[a] and B.K. Vane[b]**

[a]Department of Foods and Nutrition, University of Manitoba, Winnipeg, MB R3T 2N2, Canada; and [b]formerly Grain Research Laboratory, Canadian Grain Commission, Winnipeg, MB R3C 2G8, Canada.

## Introduction

Margarine is the earliest and most successful of the designed food staples. In the United States, per capita consumption of margarine surpassed that of butter in 1957, and within 20 years it increased to more than double the consumption figures of its prototype. As the leading tablespread in North America (1) it is now recognized as a food in its own right rather than simply as a less expensive substitute for butter.

Spreadability, particularly of the soft, tub form is margarine's unique functional advantage. However, there remains some ambivalence regarding margarine's flavor equity with butter. These two important and complex properties, spreadability and flavor are, as yet, most reliably evaluated by sensory means despite advances in rheological techniques and flavor chemistry. Accordingly, the literature related to the sensory measurement of margarine merits critical review to ensure that appropriate methods are used in research, product development, and quality control.

The primary focus of this chapter will be on the sensory methods for assessing margarine qualities as a tablespread, although the product often is used in baking and in pan frying. Special attention will be paid to sample-handling techniques during testing by both technical personnel and the group of sensory panelists who serve as the measuring instrument.

## Definition of Margarine

The worldwide standard for margarine, set by the Codex Alimentarius Commission, is a semisolid product consisting of a minimum of 80% fat, a maximum of 16% water, and with various additives being permitted, including vitamins, flavorings, colorants, emulsifiers, and preservatives (2). Physically, margarine is a water-in-oil emulsion, where the oil phase consists of both liquid oil and crystalline fat at room temperature. The semisolid structure results from a three-dimensional, continuous, sheetlike matrix of fat crystals or crystal aggregates which entraps tiny water droplets suspended in oil.

The significant sensory characteristics of margarine as a tablespread include its appearance and texture at room temperature; its melting properties in response to the modest work of spreading as well as to mouth manipulation and temperature (37°C);

**TABLE 5.1**
Consumer Opinion of Margarine Attributes

| Attribute | Self-rated importance (0 = not important; 100 = very important) |
|---|---|
| General | |
| Good color | 62 |
| Good aroma | 68 |
| Good taste | 88 |
| Good texture | 82 |
| Specific | |
| Deep color | 50 |
| Light aroma | 32 |
| Sweetness | 43 |
| Saltiness | 29 |
| Buttery taste | 63 |
| Oiliness | 10 |

Adapted from Moskowitz and Jacobs (4).

its mouthfeel; the rate of flavor release in the mouth; and the intensity of specific tastants and odorants which account for the overall flavor impression. In 1982, *Consumer Reports* (3) enlisted sensory consultants to evaluate 17 different U.S. margarines, all salted, from both stick and tub types. An excellent margarine was expected to have a pale yellow color, a slight hint of dairy flavor and aroma, an appropriate texture for its type, melt readily in the mouth, and have no more than a trace of the parent oil's flavor or aroma.

Table 5.1 shows the importance to consumers of the general sensory features of margarine color, aroma, taste and texture, and of some specific product attributes (4). Taste and texture were noted as the most important general features, while buttery taste was the most important of the specific attributes considered.

## Quality Control, Grading, and Inspection

Protocols for quality control of margarine, as with other food products, evolved in parallel with the development of the sensory sciences. Over the past decade progress has been made toward integrating the principles of sensory evaluation into the processes of quality assurance and, where appropriate, to the practices of inspection and grading.

Table 5.2 shows the margarine rating system described by Dritschel in 1970 (5) which likely predates the acceptance of the sensory sciences by food technologists. It continues to be used today for "go/no go" decisions in quality control (6). The finished product begins with a score of 100 points, and points are subtracted for specific defects in each of the four major categories. The more pronounced the defect, the higher the penalty. A final score of less than 70 indicates that the product should

**TABLE 5.2**
Example of a Scoring System for Margarine Quality Control

**Margarine Quality Rating Sheet**

*Package*

| | | |
|---|---|---|
| Dents | ___ | |
| Oil spots, dirt | ___ | |
| Innerwrap defects | ___ | |
| Miscellaneous | ___ | |
| Total (maximum 7) | ___ | Subtract from 7 = __ |

*Outer Texture*

| | | |
|---|---|---|
| Slack fill | ___ | |
| Over fill | ___ | |
| Sloshed | ___ | |
| Product adhesive to lid, wrap | ___ | |
| Oil separation | ___ | |
| Dull sheen | ___ | |
| Discoloration | ___ | |
| Blistered, grainy | ___ | |
| Total (maximum 31) | ___ | Subtract from 31 = __ |

*Inner Texture*

| | | |
|---|---|---|
| Oil separation | ___ | |
| Water separation | ___ | |
| Streaked | ___ | |
| Grainy | ___ | |
| Vaselated | ___ | |
| Cheesy, porous | ___ | |
| Total (maximum 31) | ___ | Subtract from 31 = __ |

*Flavor*

| | | |
|---|---|---|
| Lack of flavor | ___ | |
| Off oil flavor | ___ | |
| 1. Oxidized, rancid | ___ | |
| 2. Characteristic of oil used | ___ | |
| 3. Faulty oil processing | ___ | |
| Sour | ___ | |
| Plastic | ___ | |
| Fruity, artificial | ___ | |
| Old, aged flavor | ___ | |
| Total (maximum 31) | ___ | Subtract from 31 = __ |

TOTAL GRADE OF SAMPLE __

**Scoring Guides**

INNER TEXTURE

Excellent (no criticism) 31

Subtract points as follows for defects (to a maximum of 31)

Inner Texture Criticisms

| | Intensity of Defect | | |
|---|---|---|---|
| | Slight | Moderate | Pronounced |
| Oil separation | –10 | –20 | –31 |
| Water separation | –10 | –20 | –31 |
| Streaked | –1 | –3 | –5 |
| Grainy | –1 | –3 | –5 |
| Vaselated | –1 | –3 | –5 |
| Cheesy, porous | –1 | –5 | –10 |

FLAVOR

Excellent (no criticism) 31

Subtract points as follows for defects (to a maximum of 31)

Flavor Criticisms

| | Intensity of Defect | | |
|---|---|---|---|
| | Slight | Moderate | Pronounced |
| Lack of flavor | –5 | –10 | –15 |
| Off oil flavor | | | |
| 1. Oxidized, rancid | –10 | –20 | –31 |
| 2. Characteristic of oil used | –1 | –5 | –10 |
| 3. Faulty oil processing (raw, beany, grassy, burnt, fishy) | –5 | –15 | –31 |
| Sour | –10 | –20 | –31 |
| Plastic | –10 | –20 | –31 |
| Fruity, artificial | –5 | –10 | –15 |
| Oil, aged flavor | –5 | –10 | –15 |
| Grainy, sandy | –5 | –15 | –31 |

Adapted from Dritschel (5).

be withheld from market distribution. In this system the sensory characteristics of taste, odor, and texture are grouped together under the category of "flavor," a generalization of sensory descriptors which is commonly made by consumers. Defects which are sufficient to recall the product include pronounced levels of an oxidized flavor; raw, beany, grassy, burnt, fishy, sour, or plastic flavors; and a grainy mouthfeel. Other defects, not cited, may be added to the list; an example is the mouthfeel defect of chewiness or waxiness.

A limitation of this rating system for quality control is its dependence on experienced judges who have common standards of quality. A corollary is the need for adequate in-house training of new personnel and quality control apprentices. Correct identification and agreement of judges on characteristic defects is critical to the credibility of this system of quality control, as is the uniformity of their judgments over time, in terms of repeatability and validity. Similar concerns have been recognized by the government agencies responsible for grading and inspecting other foods. In Canada, this has been addressed for fish products by establishing, under the auspices of the Department of Fisheries and Oceans, a Center of Expertise for Sensory Science which is responsible for delivering university-level training in sensory analysis to all inspection personnel who evaluate fish and fish products (7). Other objectives of the Center include developing a procedures manual and establishing an odor and flavor reference-sample program to monitor inspector consistency. Such practices could be adopted to advantage by margarine associations.

Reference standards help in training and performance monitoring of quality control personnel. For example, Agriculture Canada has developed a set of standards to illustrate five specific flavor defects in butter and cheese for use by Canadian dairy graders (8). Table 5.3 describes how these standard references were prepared using whole milk as the base. Reference accuracy was established by a panel of three experienced federal dairy graders and a representative of the Food Production and Inspection Branch of Agriculture Canada. The distinction among grades for dilutions of three of the references, oxidized, sour, and bitter, was confirmed by a nine-member trained laboratory panel. Since some of the same defects found in butter and cheese may occur in margarine, these references could be useful to the margarine industry for quality control.

In Poland, Barylko-Pikielna and her co-workers (9) have argued that the industry would benefit from a more detailed sensory analysis of margarines than is required by the standard system of inspection. They used an 18–20 member sensory panel to assess nine Polish margarines using two different methods: the Polish Trade Standard (BN-80/8050-02), and intensity measurements of specific flavor attributes using six-point category scales. The data in Figure 5.1 show the scores of four attributes for each product along with the overall flavor judgment by the Trade Standard. Examining the attribute scores in detail suggests that the Trade Standard may not have classified all products appropriately. For example, product E was less buttery and had a stronger margarine/hydrogenated flavor score than others which were assigned to the highest quality category of the Trade Standard. Later work by

**TABLE 5.3**
Procedures for Preparing Flavor Defect Standards

| Flavor | Perceived Intensity | Treatment of Homogenized Whole Milk | Grade |
|---|---|---|---|
| Oxidized | Slight | 1 h, 200 Lux at 5°C | Canada 1[a] |
| | Medium | 3 h, 200 Lux at 5°C | Canada 2 |
| | Intense | 24 h, 200 Lux at 5°C | Canada 3 |
| Sour | Slight | 0.08% (w/v) lactic acid | Canada 1 |
| | Medium | 0.16% (w/v) lactic acid | Canada 2 |
| | Intense | 0.24% (w/v) lactic acid | Canada 3 |
| Bitter | Slight | 0.4% (w/v) L-isoleucine/L-leucine | Canada 1 |
| | Medium | 0.6% (w/v) L-isoleucine/L-leucine | Canada 2 |
| | Intense | 0.8% (w/v) L-isoleucine/L-leucine | Canada 3 |
| Rancid | Slight | 0.2% (w/v) lipase, 20 min at 35°C | Canada 1 |
| | Medium | 0.2% (w/v) lipase, 60 min at 35°C | Canada 2 |
| | Intense | 0.2% (w/v) lipase, 90 min at 35°C | Canada 3 |
| Fruity | Medium | 0.2% (v/v) mixture 1 with 0.04% (v/v) mixture 2[b] | Canada 2 |

[a]Maximum accepted level of flavor defect to obtain Canada 1, 2, or 3 grade if all other graded aspects are Canada 1, 2, or 3.
[b]Mixture 1: 35 μL ethyl butyrate and 50 μL ethyl hexanoate up to 100 mL in ethanol.
Mixture 2: 300 μL 1-butanol, 500 μL 2-methyl butanol, 300 μL 1-hexanol, 50 μL ethyl butyrate, 1 mL *trans*-2-hexanal, and 50 μL methyl butyrate up to 100 mL ethanol.
Adapted from Mackie and Elsaesser (8).

these investigators on the qualitative sorting of margarines from different European countries has supported the value of sensory analyses to yield detailed product profiles (10).

Tapping the talent of product specialists can enhance the work of sensory scientists. For example, Urbach and her colleagues (11) have worked effectively with a panel of professional dairy graders to establish the relevance of some lower boiling point volatile compounds to the flavor of Australian sweet cream butter. The graders' experience was particularly helpful in providing qualitative descriptions of the compounds at different concentrations.

## Sensory Methods

The application of sensory measurement can range from trouble-shooting in the market place to fundamental laboratory research. Regardless, quantifying the human response to ideas, products, or ingredients represents the application of behavioral science principles, such as perception, learning, and cognition. Therefore sensory measurements involve psychophysics in implementation and psychometrics in the design of experiments and in data analysis. The challenge for the sensory analyst is to choose or develop the most appropriate methods to address a particular question or application.

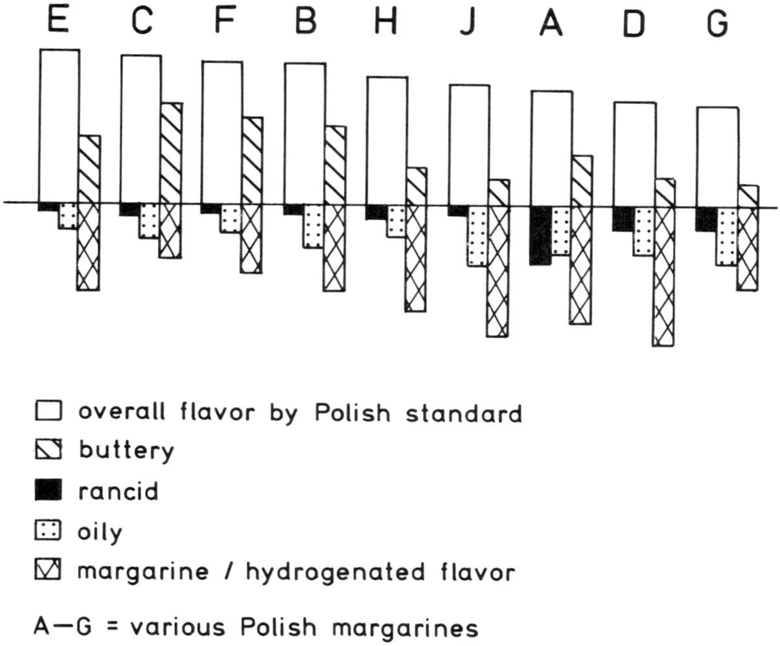


**Figure 5.1.** Comparison of margarine scores using the Polish Trade Standard vs. attribute category scales (adapted from Barylko-Pikielna *et al.* [9]).

### *Analytical and Affective Methods*

Table 5.4 presents examples of various sensory tests which have been used in the study of margarine. They have been arranged to illustrate analytical methods which will discriminate among products, describe quality attributes of a product in quantitative terms, and to show an affective method which measures satisfaction or acceptance. Analytical methods are typically used in the laboratory with screened and trained panelists. In contrast, affective methods such as hedonic scales can be used to measure the response of naive consumers. The studies cited in Table 4 will be referred to in more detail in subsequent sections of this chapter.

### *Multivariate Statistical Techniques*

In contemporary sensory research, multivariate statistical techniques can be used to better classify or define sensory attributes in a product group, differentiate among products, or predict quality. These techniques require appropriate experimental designs where the data are gathered by one or more of the analytical sensory methods listed in Table 5.4. The power of multivariate methods is illustrated in several studies involving margarines.

**TABLE 5.4**
Examples of Sensory Methods Used in Evaluating Margarine

| Function—Method | | Margarine Application |
|---|---|---|
| 1. ANALYTICAL—Discriminative: | | |
| Difference: | Measures difference or similarity between products, or attributes of products | |
| Paired comparison | | Difference in buttery flavor among blends; Fyfe and Vaisey-Genser (53) |
| Rating or scaling | | Multidimensional scaling of the difference/ similarity of tablespreads to give distinguishing features; Tuorila *et al.* (12) |
| Sensitivity: | Measures ability of individuals to detect sensory characteristics | |
| Threshold | | Least amount of $C_4$–$C_{12}$ free fatty acids distinguishable in butter by 50% of judgments; McDaniel *et al.* (51) |
| Difference threshold | | Estimation of JND from Weber ratios established for firmness and spreadability in relation to sectility and yield values respectively; Rohm and Raaber (34) |
| Power function | | Estimation of detectable differences in tablespread saltiness from the relationship between perceived saltiness and salt increments in water vs. butter; Fyfe and Vaisey-Genser (53) |
| Duration: | Measures extent and persistence of flavor | |
| Time intensity | | Effects of "margarine" melting point on diacetyl flavor intensity over time in the mouth; Lee (26) |
| 2. ANALYTICAL—Descriptive: Measures quality and/or quantity of sensory characteristics of a product | | |
| Attribute rating: | Measures intensity of a single feature | |
| Category scaling | | Degree of spreadability of margarines at different temperatures; deMan *et al.* (25); Board *et al.* (38) |
| Line scaling | | 15-cm lines anchored with descriptive terms at both ends to measure hardness and spreadability; Dixon and Parekh (40); Pokorny *et al.* (41) |

(continued)

**TABLE 5.4** (continued)
Examples of Sensory Methods Used in Evaluating Margarine

| Function—Method | Margarine Application |
|---|---|
| Ratio scaling | Magnitude estimates used to establish power functions of hardness and spreadability in relation to instrumental measures; Rohm and Ulberth (32) |
| Descriptive analysis: | Measures the intensity of several features and may describe the order of their appearance |
| Flavor/texture profile | Principal component analysis of Polish margarines using 6-point category scales; Barylko-Pikielna *et al.* (10) |
| 3. AFFECTIVE: Measures preference and/or acceptance of product | |
| Hedonic category scale | Monitoring the changes during storage in polyunsaturated margarine using a scale with nine like/dislike categories; McBride and Richardson (39) |
| Hedonic line scale | Untrained panelists assessed spreadability on an unstructured line scale anchored with hedonic faces at both ends; Rohm and Raaber (35) |

Tuorila *et al.* (12) used multidimensional scaling (MDS) to differentiate between 10 Finnish tablespreads varying in fat content (40, 60, or 80%) and in fat source (butter, vegetable oil, and blends). Differences between all possible pairs of products were measured by 18 panelists who rated overall similarity/dissimilarity on 10-cm line scales while taking into account, but not rating, the attributes of appearance, odor, texture, and flavor. Figure 2(a) is a perceptual map showing that nonparametric MDS analyses (ALSCAL) divided the products into three groups based on the first two plotted dimensions. Products 9 and 10 contained 80% fat, mainly of dairy origin; products 2, 3, and 8 contained only 40% total fat; and the remaining five products all contained fat mainly of vegetable origin, although the amount of total fat varied. Products 6 and 7 were 80% fat, products 4 and 5 contained 60% fat, while product 1 was 40% fat.

The authors characterized the first dimension as one which sorted on the basis of spreadability and butter-like flavor and were low and high, respectively, for the dairy-based spreads. The second dimension was labeled as low fat/normal fat. The third dimension (not shown in Figure 5.2[a]) appeared to distinguish the products on the melting rate. This study shows that MDS is helpful in sorting products based on perceptual dimensions underlying the sensory appreciation of a product. For the panelist, similarity ratings are straightforward, because they avoid the semantic misinterpretation inherent in attribute definitions. However, the overall task for

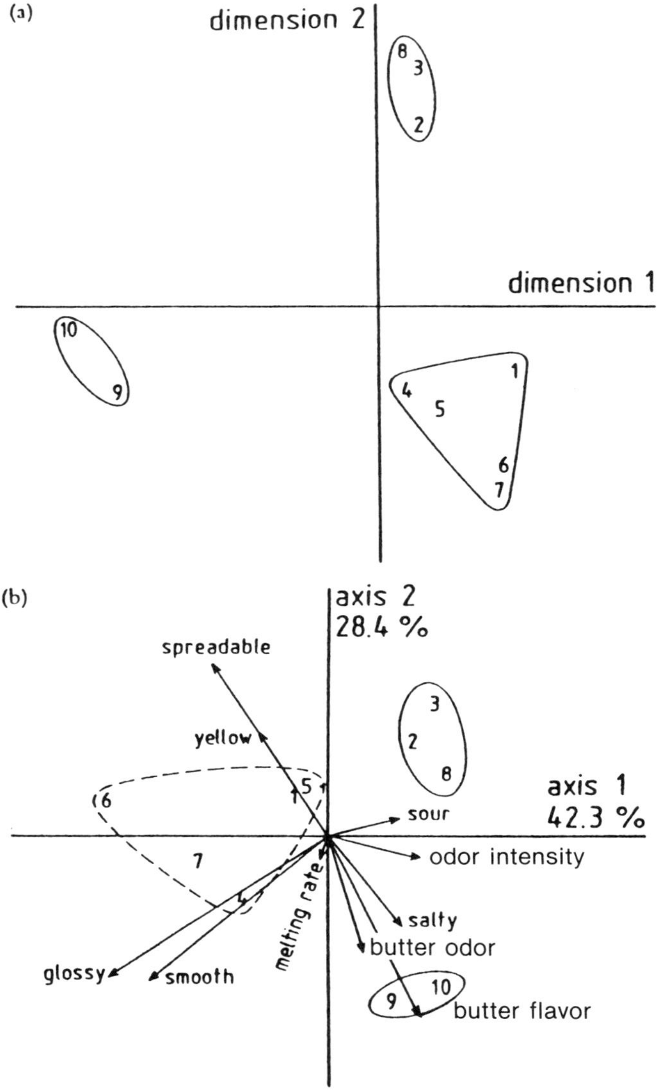


**Figure 5.2.** Similarity ratings vs. profiling of 10 tablespreads; from Matuszewska *et al.* (13). (a) MDS arrangement based on similarity ratings within pairs; (b) PCA arrangement based on line scaling of 10 attributes.

panelists may be onerous due to the need to compare all possible pairs of products. Description of the dimensions which segregate product groups must be deduced by the researchers.

Principal components analysis (PCA), like MDS, permits the arrangement of products in a two-dimensional space. However, PCA also permits the concurrent display of vectors for each attribute. The slope and length of each vector provides guidance as to its importance to the product. The relative merits of PCA and MDS for describing tablespread qualities and differentiating between products have been compared in a collaborative study between Polish and Finnish sensory scientists (13), using attribute intensity scores and similarity ratings from Tuorila's 1989 study of 10 Finnish spreads (12). The same 18 panelists who made the similarity/dissimilarity judgments which were required for MDS also judged the intensity of 10 attributes on unstructured 10-cm line scales, providing the profile data necessary for PCA.

Comparing Figure 5.2(a) with Figure 5.2(b) shows that MDS/similarity ratings and PCA/profile scores distinguished similar product groupings. However, PCA/profiling provided more detailed information about specific sensory attributes which researchers chose to have the panelists assess. Figure 5.2(b) depicts the variation between attributes and their correlation. The length of the vectors shows that glossiness, smoothness, butter flavor, and spreadability were subject to the greatest variation.

From a logistics standpoint, PCA has the advantage of efficiency since scoring a series of attributes accommodates more samples with less preparation effort than is needed for the pairwise comparisons required in similarity ratings. For example, in the study by Tuorila's group (12), 48 pairs (10 samples plus three different pairs of identical samples) had to be set up for each of the 18 panelists to produce the data for MDS analysis. This took six sessions with eight pairs per session, each pair being judged for the degree of similarity/dissimilarity on a 10-cm line scale of the overall impression of the samples, taking into account appearance, odor, texture, and flavor. In contrast, collecting the intensity scores on all 10 attributes for PCA analysis took only two sessions, where each panelist dealt with five attributes per session for each of the 10 spreads, judging intensity on 10-cm line scales. Either way, the sensory tasks were onerous. No estimates of panelist variability were reported, nor was the experiment replicated.

Multiple regression is a multivariate technique which offers some predictive power as opposed to the classification and descriptive information from MDS and PCA. Through multiple regression analysis, Vane (14) was able to show that 97% of the variability in spreadability scores, assigned by a trained sensory panel to an assortment of Canadian margarines, could be explained by a combined estimate of firmness as measured by the Texturometer, mouthmelt time as determined by the sensory panel, and reduction in firmness, also from the Texturometer. Of these three attributes, Texturometer firmness was the primary factor explaining by itself 94% of the variability in spreadability scores. Nevertheless, firmness as estimated by the sensory panel explained 93% of spreadability scores.

## Appearance

Apart from the textural features of integrity, shape, firmness, and smoothness, the appearance of margarine is influenced by its gloss and color. Both of these features are measured visually, usually by matching to some preestablished standard. In quality control, both features are judged in terms of the degree to which they are defective, for example, the extent of dull sheen or discoloration. Consumers have declared that the color of margarine is somewhat less important than taste or texture (see Table 5.1).

### Glossiness

Margarine gloss or sheen as opposed to a dull, matte surface is a function of light reflectance from subsurface layers, which is influenced by the tightness of the emulsion. Soft, tub margarines are glossier than the brick or stick forms due to the higher proportion of liquid oil. Excessive glossiness develops if the fat-crystal network is not fine enough to enmesh the liquid oil (15). Tuorila *et al.* (12) reported lower scores for glossiness in low fat than in normal fat tablespreads. The intensity of gloss was measured by 18 panelists on a 10-cm line scale anchored on either end with the words weak and strong.

### Color

The color of margarine may vary with regional preference or legislation. The model is the yellow color of butter, which varies with the breed and feed of the producing cattle. Yellow color is normally more intense when cows are on green pasturage during the summer, and is paler when they are grain-fed. Color constancy in butter over seasons may be maintained by color additives where they are permitted.

During its first century of use, margarine was often required to be distinguishable from butter by color. Legislated color novelties in margarine have ranged from no added color to pink, orange, and brown, depending on the strength of opposition to its sale by local dairy producers. In most of Europe and North America such punitive legislation had been discontinued by 1969 (16). In 1994, Canada was unique in this respect with two of its ten provinces still requiring that margarine be colored to be distinctly different from butter. The entire Lovibond tintometer scale ranges from 1.6 to 10.5 degrees of yellow and of yellow and red combined, of which 3.5 to 6.7 are reserved, in the province of Ontario, for butter (17). Beta-carotene, annatto, and tumeric are among the color additives permitted in Canadian margarines.

The color of margarine is commonly judged by one or more persons in comparison to a series of color standards. For instance, in the United Kingdom, a color scale has been devised by which butter may be matched using a disc #4/34 with 9 colors in the Lovibond tintometer. Alternately the three color dimensions, hue, value, and chroma, in a margarine may be described by comparing it visually to chips in the matte or glossy edition of the *Munsell Book of Colors* (18). Here hue represents the color identity, value the lightness or brightness, and chroma the vividness or saturation. The American Society for Testing and Materials (ASTM) has published detailed standards to guide these evaluations (19). It is imperative that the observers have normal color vision, be adapted to daylight, and view specimens which are illuminated by a controlled source against a medium gray to white background.

## Texture

The textural parameters of tablespreads, spreadability and mouthmelt, are both dependent on the relative amount of solid fat present and its crystalline structure. These determine the collective response to temperature and work applied by external forces. For a product to be spreadable, it must be plastic across a wide temperature range, but a desirable mouthmelt sensation requires that the solid fraction melt quickly and completely.

### *Solid Fat Index and Solid Fat Content*

The standard quality control assessment used by the industry on a margarine oil to predict its textural properties has been the Solid Fat Index (SFI), and has been measured by dilatometry since the 1920s (20). The method is based on the change in volume as the margarine oil changes from solid to liquid (21). The solids values are normally calculated from measurements made at three temperatures, 10, 21, and 33.5°C. More recently nuclear magnetic resonance (NMR), which depends on differences in the magnetic environment of protons in the solid and liquid phase, is being used to obtain the same type of information (15). The result, called Solid Fat Content (SFC), has been found to be more reliable, especially at high solids contents, for example, at lower temperatures (22).

Measures of the percentage of solids at any given temperature are not always as predictive of physical properties as one would like. This is due partly to the nature of the tests themselves. Because they are performed on anhydrous samples, they may not reflect how the fat will behave in an emulsion. Also, they use a standardized tempering that may not indicate the final crystalline structure of the product after manufacture and storage. Tempering during manufacture is adapted for each product, but SFI is standardized and universal. DeMan *et al.* (23) showed how different cooling and tempering resulted in different SFC. Later, they proposed tempering procedures better reflecting the end use properties of margarine (24).

In Figure 5.3, SFI values for butter and three margarines (15) have been plotted to illustrate different melt patterns as temperature increases. The sensory benchmark for such values was established by deMan *et al.* (25) who compared sensory panel assessments of spreadability with the SFI of tablespreads available on the Canadian retail market. The panel found that products with SFI values between 10 and 20, at the temperature of use, were ideally spreadable, the average SFI of products judged too hard to spread was 34.7, while a product with an SFI of 7.1 was considered too soft to spread. In light of these findings it is obvious that the stick margarine described in Figure 3, like butter, would only be spreadable as it approached 20°C. On the other hand, both the stick and tub soft margarines would be spreadable at 10°C and at higher temperatures, up to about 27°C.

### *Mouthmelt*

A margarine which melts rapidly in the mouth enjoys two sensory advantages. The flavor release is prompt and there is a subtle cooling sensation, due to the absorp-

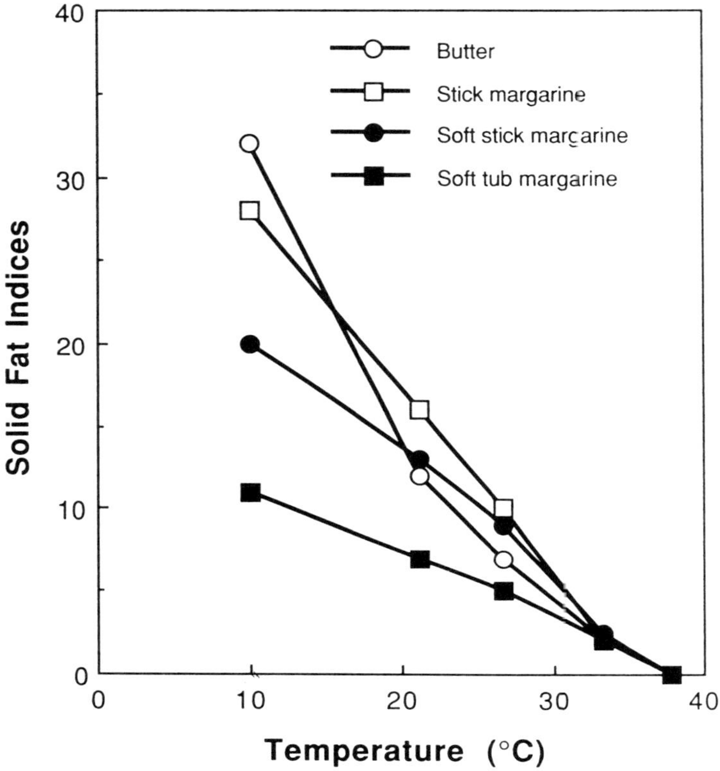


**Figure 5.3.** Solid fat indice or selected tablespreads; from Vaisey-Genser and Carr (1).

tion of the heat of crystallization in melting (15). Figure 5.3 shows that butter has a steeper SFI slope than any of the margarines which provides it with a mouthmelting advantage. This is difficult to duplicate in margarine without sacrificing the ease of spreadability inherent in products which have a low solids content at 10°C. Lee (26) has shown that the flavor impact of diacteyl was faster and greater when the base fat was palm olein than when it was palm stearin, which has a higher melting point. Margarines which fail to melt rapidly and completely in the mouth may be criticized as gummy or waxy. Chrysam (15) has stated that to avoid these defects, the product should be completely melted at body temperature and contain less than 3.5% solid fat at 33.5°C. However, this criterion makes no allowance for the time it takes for the product to melt in the mouth.

Vane (14) reported the time in seconds required for 1 mL samples of margarine to melt in the mouth, as measured with stopwatches by a sensory panel. Melt times for both brick and tub forms of margarine were clocked when the temperature at the time of introduction to the mouth was either 4 or 21°C. The data in Table 5.5 show

**TABLE 5.5**
Average Mouthmelt Time of Selected Margarines

| Market Form | No. of Brands | No. of Sensory Observations (7 panelists x 2 replications) | Mouthmelt time (sec) from 4°C[a] $\overline{X}$ (brand range) | Mouthmelt time (sec) from 21°C[b] $\overline{X}$ (brand range) |
|---|---|---|---|---|
| Margarine | | | | |
| Brick[c] | 6 | 84 | 22.4 (20.0–25.1) | 15.8 (12.9–18.2) |
| Tub[d] | 5 | 70 | 13.6 (12.3–14.4) | 9.4 (6.0–11.0) |
| Butter | 1 | 14 | 18.2 | 11.0 |

[ab]Difference between temperatures significant at $P < 0.001$.
[cd]Difference between market forms significant at $P < 0.05$.
Source: Vane (14).

that the mouthmelt time was significantly longer for products sampled at refrigerated temperatures than when they were at room temperature, and that soft tub margarines melted more rapidly than the firmer bricks. The mouthmelt time for butter was shorter than brick margarines under either temperature condition, for example, 18 sec vs. 20–25 sec at 4°C and 11 sec vs. 13–18 sec at 21°C. The margarines in this experiment were selected to represent the range of oil sources in Canadian retail products. They were obtained directly from the manufacturers to assure comparable processing age and that both market forms within an oil source/brand were produced under common conditions.

Vane's panel (14) consisted of seven women selected for their ability to recognize textural differences in deliberately aged margarines. They were then trained to accept common definitions of the attributes which they decided were important to measure. The definition for "rate of mouthmelt" along with the instructions the panelists elected to follow in making the measurement are shown in Table 5.6. Samples for mouthmelt evaluations were measured onto coded plastic spoons and equilibrated at 4°C overnight. Samples to be tested at 21°C were held for 30 min at room temperature prior to serving, whereas the 4°C samples were served on ice trays.

## *Graininess*

Graininess is a defect in margarine which can develop if the solid fat crystals pack together to form particles sufficient in size and number to be perceptible (27). Vane (14) had a sensory panel make two graininess judgments, size and amount, on several margarines, using line scales (see Table 5.6). Graininess references for particle size were a starch slurry, canned pears, and cream of wheat, which were used during training to familiarize the panelists with the continuum of chalky, gritty, and grainy, respectively (28). The decision to do the two evaluations sequentially on the same sample was made during panel training. Because the second judgment was dependent upon the first, this practice did not introduce the prejudice inherent in

**TABLE 5.6**
Instructions for the Sensory Assessments of Tablespread Texture

1. Rate of mouthmelt: The time required for a product to melt completely in the mouth[a].
Take the entire sample into your mouth. Move the sample from side to side across the palate with your tongue, being sure to apply the same pressure to each sample. With a stopwatch, observe the time required to achieve melt. Record this time in the space provided.

| Sample | Time to Melt (Seconds) |
|---|---|
| ______ | ___________ |

2. Graininess: The presence of discrete particles in a product[b].
There will be two judgments, the extent of graininess and the size of the particles if present. Lick the sample into your mouth. On the first line for each sample, indicate with a vertical mark the assessment of the quantity of particles present. For those samples with particles, please indicate the size of the particles by marking the second line.

no particles ________________________ many particles

chalky ________________________ grainy

3. Firmness: The force required to obtain a given deformation[c].
Compress the samples with the spoon handle. Score each sample on the magnitude of resistance to compression, relative to the first sample provided, which will serve as the reference.

| Sample | Score |
|---|---|
| ______ | _____ |

4. Work to soften: The resistance of a product to structural breakdown[c].
(R = 60% vaseline; 40% cornstarch, wt %).
Feel the consistency of the reference sample with the spoon handle. Assign the reference a "work to softening" score. Work each of the samples with the spoon handle until they equal the softness of the reference. Assign scores relative to the amount of effort required to effect the desired consistency change.

| Sample | Score |
|---|---|
| ______ | ____ |

5. Spreadability: The ease with which the sample can be applied in a thin even layer to bread[c].
Spread one-half the sample on the bread with the knife provided. Assign a spreadability score to each sample relative to the first sample. The more easily spread products receive a higher score. The remaining one-half of the sample can be used to confirm the initial impression.

| Sample | Score |
|---|---|
| ______ | _____ |

Using the numerical scale used previously, assign scores which correspond to the following (see Table 5.7).

[a]Sample = 1 mL measure.
[b]Sample = approximately 5 mL leveled on a plastic spoon.
[c]Sample = 2-cm diameter ball.
Adapted from Vane (14).

requesting assessments of two independent parameters on the same sample. Graininess was more likely to occur in brick type products and at refrigerator temperatures.

Vaisey-Genser *et al.* (29) used a seven-member panel to assess the presence or absence of graininess in margarines at 2-week intervals during storage for 6 months at 4°C. As an example of graininess, panelists were provided with a blend of cooked farina in margarine (30 mL in 57 g). Crystal size was measured from photomicrographs taken under polarized light. By plotting percent detection against crystal size, the sensory threshold for perceived graininess of margarine crystals was estimated as 22μ. Earlier, Vane (14) had observed that graininess perception appeared to be a function of the number of crystals, as well as their size.

Hough *et al.* (30) clarified the relationship between crystal size and concentration in studies on lactose-induced sandiness perceived in *dulce de leche,* an Argentine dairy product similar to sweetened condensed milk. They established that the power function relating perceived sandiness to crystal size was ≈ 1, and proceeded to develop a seven-point sensory scale for sandiness using *dulce de leche* mixed with sand of crystal sizes varying between 0 and 1,500 μ. A 12-member panel measured perceived sandiness of a series of *dulce la leche* products in which crystal size was controlled, but crystal quantity varied. Sandiness was not detected with a crystal size of 6 μ. The detection threshold for sandiness ranged from 45–105 μ, with the smaller size being perceived only at higher crystal concentrations.

### *Firmness vs. Spreadability*

Firmness as a discrete sensory property of a tablespread is implicit in the determination of its spreadability (22). The first impression of a tablespread texture will be that of its firmness, obtained as the consumer forces a spreading implement into the product. Spreadability, in pragmatic terms, is the ease with which a spread can be applied in a thin, even layer to bread (14). Firmness or hardness may be measured by the force required to obtain a given deformation or by the amount of deformation under a given force. Although spreadability is also a deformation under an external load, it is a more dynamic property (22). Measurements of firmness and spreadability are usually highly correlated, however the relationship is rarely perfect, and this is partly a function of worksoftening. Margarine worksoftens more easily than butter, which allows it to be more spreadable even when hardness values are initially equal (31). Sectility tests have been shown to correspond to sensory estimates of hardness, while the apparent yield value calculated from cone penetration with constant load correlated best with sensory estimates of spreadability (32).

Sensory assessments of margarine spreadability have been done from as simply as a "great many thumb tests" (33) to as precisely as a study of difference thresholds (34). In the early work of Haighton (33), no details of the sensory methods were given, but judgments were made about spreadability in relation to the results of cone penetration tests. Yield values in the range of 100 to 1,500 g/cm$^2$, at any specific temperature were considered descriptive of a spreadable product.

Subsequent research has narrowed this range. In 1979, deMan *et al.* (25) reported that margarines having a yield value in the range of 125–500 g/cm$^2$ generally received favorable spreadability ratings when assessed by a 6–10 member sensory panel. In 1991, Rohm and Raaber (35), using 33–40 untrained panelists, established hedonic spreadability optima in the apparent yield value range of approximately 30–60 kPa, which, using a factor of 10.197, converts to 300–600 g/cm$^2$.

More research appears to have been published on the firmness and spreadability of tablespreads than on any other sensory property. This wealth of literature affords the opportunity to examine the efficacy of the various sensory methods which have been used in the effort to relate sensory perception to physical measurements of margarine.

*Category Scales.* Category scales are commonly used for assessments of more than two products. Panelists are not requested to compare the samples directly; rather each is judged on its own merit according to the category labels. The labels may be presented on the sensory ballot as a simple list or as anchors on a line. Numeric values are normally assigned by the experimenter after the panelists have indicated their category of choice. For the effective use of category scales, all panelists should agree in advance on the meaning of the scale labels; for example, "hard" should require the same amount of pressure for each panelist. The noncontinuous nature of category scales makes nonparametric statistics appropriate for investigating differences among samples.

Analyses of data accrued from category scales may be flawed with the assumption that each category is equidistant. This has not always proved to be the case. In 1960, Riel (36) used a 15-member panel to assess the firmness and spreadability of tablespreads concurrently, using the following five categories: much too hard, hard but acceptable, desirable, soft but acceptable, or much too soft. Judgments were made by cutting with a knife and spreading the sample on a piece of bread. In comparing the sensory results with measurements of spreadability on the Huebner-Thomsen apparatus, Riel (36) reported a high correlation ($r = 0.96$). However, he noted that the instrumental differences were greater on the firmer side of the scale.

An examination of later work by deMan *et al.* (25) confirmed the observation of unequal distances between sensory categories of spreadability. A panel of 6–10 members judged spreadability on a nine-point category scale labeled with four levels for both hardness and softness and having a midpoint of "ideally spreadable." Judgments were made while spreading samples of 9 margarines and butter on soda crackers at test temperatures of 5, 10, 15, 20, and 25°C. From plots of hardness, determined by cone penetration vs. spreadability, it appears that the range of hardness values was greater between the sensory categories of ideally spreadable and extremely hard than it was between the ideal and extremely soft (25). Vane (14) showed further that, in the seven-point category scale for spreadability which panelists derived from magnitude estimates, the psychological difference appeared greater on the firmer side of the midpoint (Table 5.7). It seems that it is more difficult to discriminate degrees of softness than of hardness.

**TABLE 5.7**
Mean Values Assigned to Spreadability Categories

| Category | n | Mean |
|---|---|---|
| Too hard (will not spread) | 51 | $0.7^{a}$ |
| Too hard, just barely spreadable | 54 | $5.3^{b}$ |
| Hard, but satisfactorily spreadable | 58 | $7.6^{bc}$ |
| Satisfactorily plastic and spreadable | 64 | $10.0^{cd}$ |
| Soft, but spreadable | 51 | $11.5^{de}$ |
| Very soft, oil soaks into bread | 50 | $12.6^{de}$ |
| Very soft, almost pourable | 43 | $14.2^{e}$ |

$^{abcde}$Means followed by the same letter are not significantly different at $P < 0.01$
Adapted from Vane (14).

Most category scales for degrees of spreadability appear to have used the term firmness/hardness synonymously or interchangeably with spreadability. This practice remains debatable. It is defensible in Vane's work (14) where sensory and Texturometer firmness scores explained 93% and 94% of the variability in spreadability scores, respectively, but it is less so in deMan's 1979 study (25) where only 58% of the variability in spreadability scores was explained by instrumental hardness.

Consumer's Union (37) scored commercially available products on a spreadability category scale ranging from 1 = very soft, very easily spread, no resistance, and oily, to 5 = very difficult, with the arbitrary descriptor "normal" as the midpoint. The amount to cover a slice of bread was also measured and described in four categories ranging from too thin a layer (1/4 Tbsp) to too thick (1 Tbsp), providing a novel functional interpretation of spreadability.

Most researchers have elected to use a bipolar structure for category scales of spreadability. That is, the midpoint represents a degree of firmness of optimum spreadability, with categories of too hard or too soft to spread on either side (see Table 5.7). One exception is the work of Board *et al.* (38), who used a unipolar seven-point category scale ranging from 1 = extremely poor spreadability to 7 = ideal spreadability, in examining tablespreads at different temperatures. Although the 20 panelists were experienced in sensory assessment, they were not trained specifically for judging margarine spreadability. Their instructions included a reminder that products which were either too hard or too soft could receive low spreadability scores. After examining the standard errors of the data, the authors concluded that the panelists were most consistent in assessing samples at either extreme of the scale and were more variable in assessments of intermediate spreadability. Butter and brick margarines were considered optimally spreadable at around 20°C whereas tub margarines were considered to have very good or better spreadability between 0 and 10°C.

McBride and Richardson (39) used nine-point hedonic scales with a 40-member panel to assess the shelf life of a polyunsaturated margarine during storage. The nine categories ranged from like extremely to dislike extremely with neither like nor dislike as the midpoint. Panelists were given a single sample which they were asked to spread on a piece of bread, to taste, and then to give separate hedonic judgments for each of four parameters: color, flavor, texture, and general acceptability. The practice of requiring people to judge more than one parameter on the same sample is not recommended. It compromises the independence of sequential judgments, since they are susceptible to bias from previous decisions. This hazard is particularly acute when the category labels are solely hedonic.

*Line/Intensity Scales.* Line scales are more common than category scales in contemporary sensory studies. Their advantages are that they avoid decisions by the experimenter about category labels and spacing and that they are less constraining in actual use by the panelists. A line scale will be considered unstructured or structured depending on how many reference points or anchors are provided along its length. At a minimum, an anchor is needed at or near either end of the line. Line length varies, but the norm is 15 cm. Panelists can indicate their judgments by placing a mark at any point on the line and may indicate minor differences between products which may have been grouped together under a category scale. Data points are translated to linear measures for analysis which can be performed using traditional parametric procedures.

Dixon and Parekh (40) used a 15-cm line scale to measure the spreadability of several butters. The line was structured with three anchors, sloppy and hard within 1 cm of either end, with spreadable as the midpoint. They found good agreement among their four panelists in the range of spreadable to hard, but results diverged on the soft side of the scale. This suggests that placing the anchor "spreadable" at the midpoint may have inappropriately truncated the firmer side and elongated the softer side of the spreadability line. This points out how important it is that the location of anchors on line scales be determined by the panelists during the training period preceding an experiment, rather than being decided unilaterally by the researcher.

Pokorny *et al.* (41) used unstructured 15-cm line scales to measure three textural parameters of margarine: hardness by cutting a 5-cm cube with a knife, hardness by spreading on a piece of bread, and spreadability by spreading on a slice of bread. There were 405 panelists who tested 28 soft and semisoft experimental margarines in the temperature range of 5–25°C. The scales were anchored only at the ends of the lines with the terms extremely soft, extremely hard, or very well spreadable, very poorly spreadable, respectively. The fact that the panelists reported difficulty in judging the spreadability of very hard and very soft samples points to the value of prior training. The sensory data were compared to rheological measurements made with a Ferranti-Shirley cone and plate viscometer. Although the correlation coefficients were statistically significant, they were consistently lower for the softer margarines. The authors concluded that sensory analysis could not be replaced with rheometry.

Rohm and Raaber (35) also used untrained panelists to assess spreadability of six different tablespreads using an unstructured line scale. However they chose to anchor the scale at both ends with faces representing the degree of preference and, prior to the assessments, their 33–40 panelists were introduced to the nature of hedonic testing. Subsequently they converted each panelist's scores for the six products to ranks and, by relating the rank totals to apparent yield values, established hedonic spreadability optima.

*Ratio Scaling or Magnitude Estimation.* The method of magnitude estimation imposes neither the limitations of length and anchor placement that are inherent in line scales nor the constraints of label definition and number that are experienced with category scales. Panelists are instructed to consider the magnitude of difference between a reference and another sample, as a ratio. An identified reference sample may be used which has a preassigned score (fixed modulus), or it may be assigned a value of the panelist's choice (free modulus). Alternately the first coded sample in a randomized series may be scored by the panelist and used thereafter as the reference for the remaining ratio judgments (free modulus). For example, if the first sample of a margarine was assigned a value of 10 to represent its ease of spreadability and the second sample was considered 50% easier to spread, then the second sample would be assigned a value of 15. Appropriate data handling and analysis is required depending on the approach used (42).

In comparing the textural qualities of 11 retail margarines and butter, Vane (14) adopted the fixed modulus in measurements of firmness, with a reference of medium hardness, and used the free-modulus approach with the first tablespread sample as a reference for magnitude estimates of spreadability. A common vaseline-cornstarch reference was used by all panelists for estimating ratios of work softening (see Table 5.6). Sensory firmness estimates correlated well with the instrumental measurements of firmness from the Texturometer, with coefficients of $r = 0.95$ and $r = 0.99$ at 21 and 4°C, respectively. Sensory estimates of "work to soften" were limited to the samples tested at 4°C because, at room temperature, the margarine samples were softer than the vaseline-cornstarch reference. Panelists reported that a softness criterion was used when "work to soften" was not applicable. The association between sensory and Texturometer measurements of work softening ($r = 0.83$) was not as strong as that for firmness. A total of 97% of the variability in the sensory estimates of spreadability was explained by a multiple regression equation which combined instrumental measurements of firmness (94%) and sensory measurements of mouthmelt time (3%).

In order to convert spreadability ratios to functional properties, Vane (14) asked the panelists to assign ratio scores to a category scale, after they had made magnitude estimates of spreadability on coded samples (see Table 5.7). Although the conversion provided interesting information on the perceived differences among categories, this practice introduced a possible bias. Ratio scores, once assigned to a category label, could become internal reference points, such that future judgments would be category assignments instead of magnitude ratios.

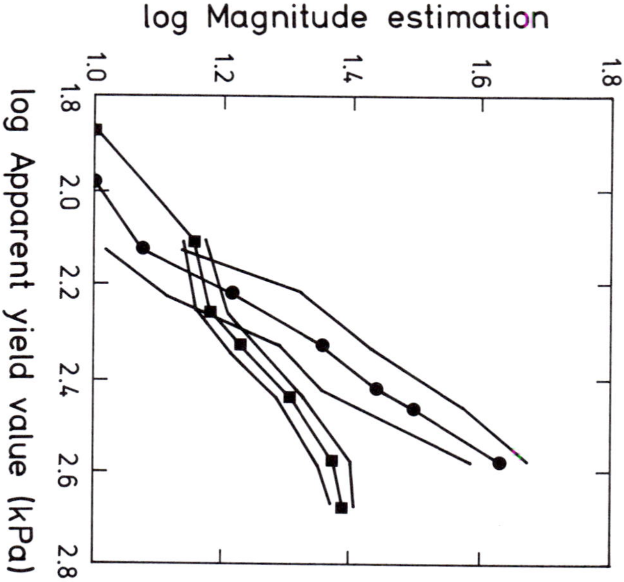


**Figure 5.4.** Effect of training on the variability in magnitude estimates of butter spreadability; ● = untrained panelist; ■ = trained panelist; from Rohm (44).

In 1989, Rohm and Ulberth (32) showed that the relationship of magnitude estimates of butter firmness (*S*) to sectility measurements (*C*) and of butter spreadability to apparent yield values each conformed to the general power law equation, $S = kC^n$, thus validating their further mathematical treatment. The slopes/exponents ($n$) of the lines were within the range of 0.4–0.7, which is customary for texture responses based on fundamental measurements (43). An exponent of less than one means that doubling the stimulus (*C*) would not double the sensory intensity (*S*).

The trained, five-member panel used reference samples with an assigned value of 10, for both hardness and spreadability assessment. Panelists were asked to assess the ease of spreading a cylindrical sample with a knife onto filter paper on a perspex plate and to judge hardness as the force required to cut the sample. Later, the same type of study was carried out using 15 untrained panelists (44). While the results still conformed to the power-law equation, the variability in spreadability estimates was greater than had been the case with trained panelists in the earlier experiment. The data compared in Figure 5.4 show the value of an appropriate training period prior to sensory analysis in reducing the variability in spreadability scores.

*Difference Thresholds.* In 1992, Rohm and Raaber (34) reported the just noticeable differences (JND) in instrumental measures of the spreadability and firmness of tablespreads which could be perceived by a 12-member panel. The linear relationship between JND and log instrumental values showed that the JND were not subjectively equal, but increased with increasing stimulus intensities, conforming to Weber's law. To estimate the JND, difference thresholds of six tablespreads were measured using a paired comparison design. A texture midpoint (TM) sample was selected for each spread. Panelists were then randomly presented with six other samples of the product which were varied in temperature to create a hardness and spreadability range that surrounded the TM. For each sample, panelists were asked to state whether or not it was softer/easier to spread than the TM. Three replicates of each pair were done by each of 12 panelists giving a total of 36 assessments for each pair, and 216 judgments for each TM for both hardness and spreadability. The percent correct was then plotted on normal probability paper against the log instrumental measures of firmness and spreadability. Results indicated that the panelists were better at judging firmness than spreadability. The demonstration of conformance to Weber's law increases the confidence with which predictions of sensory firmness and spreadability may be made from instrumental assessments.

## Flavor

In spite of the fact that North Americans use more margarine than butter, the consumer attitude persists that butter has better flavor. In 1969, 74% of 1,157 North American consumers surveyed by telephone believed that butter was "better tasting" than margarine (45). A 1986 survey of 4,000 U.S. consumers by the United Dairy Industry Association found that the majority still preferred butter (46). Sensory reports from Poland and Finland have shown that margarines often lack buttery flavor (10,12). The challenge remains for the margarine industry to mimic butter flavor better.

### *Margarine Flavor Models: Cultured vs. Sweet Cream Butter*

Matching margarine flavor to butter is complicated by regional differences in the types of butter traditionally available. North Americans are more familiar with salted and sweet cream tablespreads, while cultured butter and unsalted spreads are the norm for many continental Europeans. While both cultured and sweet cream butters have sophisticated flavors, Ramshaw (47) considered that cultured butter flavor was easier to duplicate than the more subtle and complex flavor of sweet cream butter.

Identification of the chemical compounds associated with butter flavor and aroma is a first step in its simulation. The next is to establish the appropriate quantitative balance of the flavor components by sensory evaluation. Using this approach, Lindsay and his colleagues (48) designed a synthetic butter culture flavor concentrate consisting of a combination of 250 ppm lactic acid, 30 ppm acetic acid, 0.2 ppm acetaldehyde, 0.08 ppm dimethyl disulfide, and 0.5–2.0 ppm diacetyl, adjusted to pH 5.2 with sodium bicarbonate. In developing the concentrate, starting levels of flavor compounds were guided by chemical analyses of normally cultured butter and consideration of reported sensory thresholds. Ongoing sensory evaluations by an 8–10 member panel of the desirability of flavorants when dispersed in acidified, heated, raw whole milk provided further guidance. The final synthetic concentrate, in a propylene glycol carrier, was added to churned fresh cream at the time of moisture adjustment. The product containing the synthetic flavor concentrate proved to be as well-liked by a panel of 175 university students as butter cultured in the normal manner.

Comparable success in simulating sweet cream butter flavor appears yet to be published even though the concentrations of a number of its volatile flavor components and their sensory thresholds have been reported (47). Urbach and colleagues (11) have stated that "any compound whose actual concentration-to-threshold ratio is greater than 1 makes a contribution to butter flavor." However, sensory thresholds have been found to vary with the media in which the compounds are dispersed for testing, and with the experience and sensitivity of the panelists (49,11).

### *Threshold Measurements*

Siek and co-workers (49) demonstrated the effects of different media on flavor thresholds. They reported that the threshold for diacetyl (2,3 butanedione), was almost twice as high in deodorized butteroil (0.055 ppm) than it was in butter (0.032 ppm) and ten times higher than in water (0.0054 ppm). They measured the thresholds of several butter volatiles by having 20–28 subjects test four or five dilutions of each flavorant in a specific medium and simply state whether or not a flavor was present. The concentration at which 50% of the responses would be positive was then calculated as the threshold (50). Threshold determinations usually required one or more preliminary tests to find the appropriate concentration range. During each test, subjects first examined two labeled references, a blank and a sample of the highest concentration. Both references were included as coded samples, and the data were discarded from subjects who identified them incorrectly. All butteroil samples were coded and served at 42°C in a randomized order. Subjects examined them by sniffing and/or tasting.

McDaniel *et al.* (51) measured thresholds in sweet cream butter of the $C_4$–$C_{12}$ free fatty acids (FFA), using the flavor threshold method described previously with some refinements. The results were expressed two ways; first as the amount of added FFA at threshold, and second as the sum of the latter value plus the endogenous level measured previously in the butter. Fourteen panelists were asked to compare each sample to the identified blank for flavor difference, making the test essentially one of multiple-paired comparisons. Thresholds were calculated from the linear regression of positive responses on FFA concentration and, at the 50% positive level, were referred to as "average flavor thresholds" (AFT). In preparing the samples, each FFA concentration was mixed thoroughly into the butter by kneading in a plastic bag. The flavored butter was shaped and chilled for 24 h before testing.

Urbach *et al.* (11) used synthetic butter prepared from deodorized butteroil (84%) and distilled water (16%) to examine the thresholds and qualitative features of a number of butter volatiles. Liquid flavor material, diluted as appropriate in small amounts of molten butter, was added to solid butter and dispersed with a spatula without melting the butter. Flavor evaluations were made by panels of three to six experienced professional butter graders using the open grading method which is based on free discussion among the judges. The flavor threshold was defined as the lowest concentration in synthetic butter at which more than one-half of the graders present could differentiate between a blank sample and one containing the additive. In several instances, thresholds were lower than those previously reported from sensory panels. The authors attributed this to the experience of the grader/judges. The changing composition of the panel and the fact that the studies were done over a period of 4 years, precluded statistical treatment of the data. While no single compound alone could account for butter flavor, the following compounds were considered likely to make important contributions: $C_8$, $C_{10}$ δ-lactones, $C_{10}$ acid, phenol, *p*-cresol, indole, and skatole.

### Power Functions

While flavor thresholds in butter have been considered as starting points in formulating margarine flavorings, their usefulness is limited by the fact that they are only single-point measurements. Knowing the concentration of a flavorant at the sensory threshold provides no reliable information on which to predict the impact of higher concentrations. In contrast, the power function ($S = kC^n$) shows the growth of a perceived flavor over a range of suprathreshold concentrations, so that the effect of flavorant increments can be forecast (52).

As demonstrated in Figure 5.5, Fyfe and Vaisey-Genser (53) established the power functions for salt intensity in two series of salt concentrations, one in water (5 levels) and one in butter (8 levels). The butter samples of varying salt concentration were prepared from a single lot of butter produced in a dairy pilot plant. Seven trained panelists estimated saltiness compared to a reference sample (0.3% salt solution in water) using the method of magnitude estimation (52). The slopes ($n$) of the sensation ($S$): concentration ($C$) lines, were similar in both water and butter, showing that the rate of growth of perceived saltiness was media-independent. However, the difference between the intercepts ($k$), the distance in elevation between the lines, indicated that about ten times as much salt was needed in butter to elicit a salty sensation equivalent to that of salt in water. In Figure 5.5, panelists' magnitude estimates of perceived saltiness of eight margarines and butter were plotted against the label claims for salt. In seven out of nine cases, saltiness scores fitted the salt-in-butter function reasonably well. This suggests that the power function for saltiness can be a useful tool in predicting the flavor impact of altering salt levels in margarine.

Establishing the power functions for some of the fat-soluble volatiles which have been identified as being important in butter flavor, would appear to be a promising approach to designing flavor mixtures for use in margarine. Such functions provide a basis for quantifying the relationship between concentration and sensory impact for flavor components, individually and in combination (54).

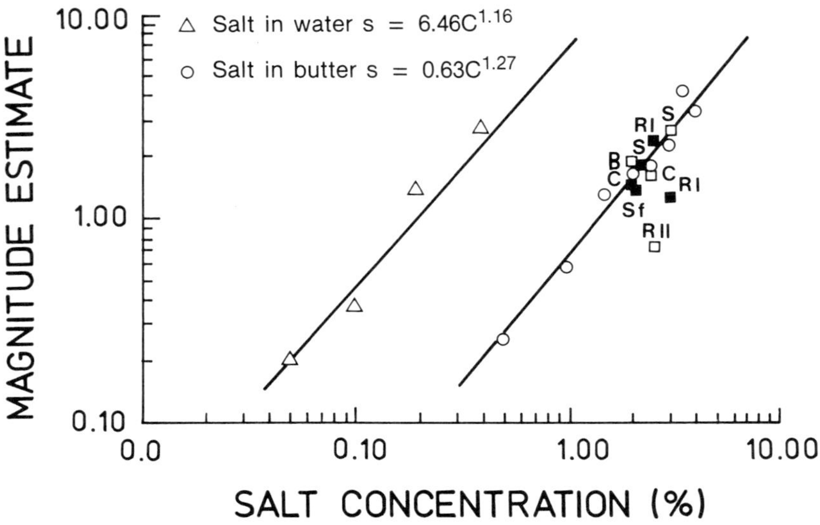


**Figure 5.5.** Relationship of salt intensity with salt concentration in water and in butter; RI, RII, C, S, and Sf = various retail margarines and B = butter; from Fyfe and Vaisey-Genser (53).

### *Multiple-Paired Comparisons*

Blending margarine with butter has been one solution to add buttery flavor while retaining the spreadability advantage of margarine. Fyfe and Vaisey-Genser (53) compared the relative buttery flavor intensity of margarine with two butter-margarine blends (50:50 and 20:80, respectively) and butter using the Scheffé multiple-paired comparisons test. This method called for all panelists to compare the four samples as all possible pairs (six pairs) with the first-sample order balanced among judges and the order of pairs randomized for each judge (55). For each pair, the judge had only to declare which sample had the stronger buttery flavor. All spreads were brick market forms. The basestock for the margarine and the blends was canola oil, which had been obtained from the same processor.

The sensory panel consisted of six women who had been trained to distinguish between butter and margarine using triangle tests and to scale, using magnitude estimation, the intensity of buttery flavor in a series of commercial margarines. The panel found that margarine had significantly less buttery flavor than butter. Of the two blends, the 50:50 product was intermediate; that is, it was neither significantly less intense in buttery flavor than butter nor significantly more intense than margarine. On the other hand, the blend which contained only 20% butter, like the 100% margarine, had significantly less buttery flavor than butter.

The multiple-paired comparison design permitted the examination of several products, while requiring only a single task of the judge for each pair. However, the validity of the paired test hinges on the absence of any difference between pair members other than the one in question. Despite efforts in this experiment to minimize textural differences among products by serving them on coded spoons and at room temperature, the possibility of mouthmelt differences influencing the distinction between samples could not be ruled out.

### Time-Intensity Measurements

The physical characteristics of both the lipid and the aqueous phases of the margarine emulsion not only determine its texture but also influence its flavor. Chrysam (15) observed that "if the aqueous droplets are uniformly small and heavily stabilized with emulsifiers, the flavor and salt release will be delayed."

Lee (26) used the time-intensity method to demonstrate that differences in the melting rate of the fat base affect the intensity of the flavor perceived in a margarine-like model system. Two simulated spreads were produced from liquid soybean oil melted with palm fat (60:40 w/w), flavored with 0.01% diacetyl and other dairy flavors, and recrystallized at 0°C. Palm olein was used in one spread, and the higher melting point palm stearin in the other. Eight trained panelists used an unstructured line scale to judge total flavor intensity of 2.5 g samples which had been equilibrated to 0°C. Panelists noted their opinions from the time that they placed the sample in their mouths to the time that the flavor intensity returned to zero after swallowing.

Figure 5.6 shows that the average total flavor intensity for both spreads peaked after 12 sec in the mouth. However, at that point and to extinction, flavor intensity was consistently greater in the more rapidly melting palm olein product. Mass spectrophotometric measurements of diacetyl, in the headspace of the same products held at 35°C in the presence of artificial saliva and agitated with glass beads, confirmed the sensory findings. This elegant study highlights the value of time-in-the-mouth guidelines for panelists being trained in the assessment of tablespread flavor or, better still, of using the time-intensity method for measuring margarine flavor.

## Shelf Life

Standard margarine has a shelf life of 6–12 months, provided that the product is refrigerated during distribution, marketing, and home storage. The period will be shorter for low-fat, high-moisture products and for those which are salt-free. Spoilage hazards include microbial growth, autoxidation, photochemical oxidation, and a crystal structure prone to polymorphism.

McBride and Richardson (39) assessed the acceptability of a polyunsaturated sunflower oil margarine during storage for 12 months at 5, 10, and −30°C. The information was needed in order to classify the product for mandatory date coding under Australian food laws. They had a 40-member untrained panel examine samples at 2 month intervals and score them on nine-point hedonic category scales for

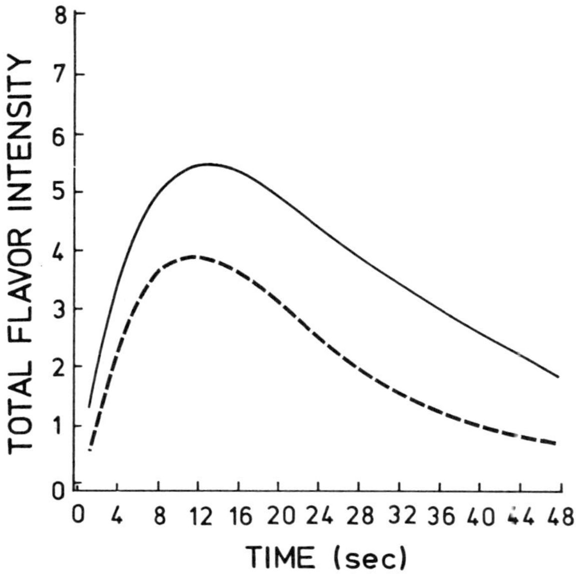


**Figure 5.6.** Effect of margarine melting point on the time and intensity of flavor release; — palm olein, --- palm stearin; from Lee (26).

flavor, color, texture, and general acceptability. Because all parameters were judged on the same sample, it is not surprising that the pattern of responses for flavor and general acceptability were similar, with significant effects due to storage time, storage condition, and their interaction. This parity suggested to the authors that flavor impression played a dominant role in the assessment of general acceptability.

From work on the shelf life of frozen foods, Gutschmidt (56) proposed that a drop of one point on a nine-point hedonic scale corresponded to the end of high-quality life for a product, and that a score of six (like slightly) on a nine-point scale represented the limit of practical storage life. Applying these criteria to their data, McBride and Richardson (39) concluded that the limit of practical storage life for the polyunsaturated margarine, when stored at 5 or 10°C, was 12 months. However, a decline from high quality was noted after 8 months at 5°C, and after 6 months at 10°C. Distinct graininess was noted in the samples exposed to −30°C and apparently accounted for the lower scores for texture acceptability.

Vaisey-Genser *et al.* (29) studied the textural stability of two canola oil margarines by monitoring the frequency with which graininess was detected during 6 months of storage at 4°C. A seven-member sensory panel examined the products at 2-week intervals. One margarine remained smooth throughout storage, while the other became perceptibly grainy after 12 weeks. Sensory testing proved effective for testing margarine formulations designed to limit the crystal growth responsible for graininess development.

## Sample Handling

Controlled handling of margarine by technical personnel during the preparation of test samples and by panelists during testing adds validity to the results. Where textural features are being studied, it is important to consider worksoftening during sample preparation and to standardize the recovery time before serving. Butter work softens less (50–55%) than margarine (70–75%) and recovers more readily, although neither recovers fully (31,57). For both products, the amount of recovery is directly related to the extent of primary bond rupture (58). Sampling practices which involve cutting the sample into cubes (37,41), butter balls (14), or cylindrical cores (34) have less impact on the internal structure of the sample than those which involve loading it into a container (38).

Temperature control of tablespread samples is an important test prerequisite. Board *et al.* (38) studied spreadability at six temperatures between 0 and 25°C. Just before testing, the dishes were moved to the testing area in insulated boxes which contained plastic bags of water at the test temperature. Thermocouples in the samples showed changes of less than 0.5°C before the panel received them. The serving dishes were placed in a block of polystyrene foam and covered with foil which limited the temperature change to less than 2°C within 10 min; panelists usually completed their sensory assessments in less than 5 min. Vane (14) controlled samples at 4°C by arranging the samples for each panelist on an aluminum tray in a refrigerator adjacent to the testing area. Panelists picked up an ice pack from the freezer and placed the aluminum tray directly on it, which kept the samples from warming during the test period. Rohm and Raaber (34) conditioned all samples overnight in thermostatically controlled water baths; sensory analysis was completed within 1 min of removal from the unit.

The receptor used for spreadability measurements has most often been a slice of bread (14,36,39,41). A less variable carrier is preferred. Examples include soda crackers (25) or a perspex plate covered with filter paper (34).

In studying margarine flavor formulations, exploratory work may be carried out with heated milk as Siek *et al.* (49) did in simulating cultured butter flavor. However, preliminary testing in a "home-made" margarine may be more efficient. A recipe for such a product is shown in Appendix A. Alternately, a commercially made unflavored margarine may be used as the base for testing flavor additives. When this is the case, flavorants may be worked in with a spatula (11) or by kneading in a plastic bag (51).

## Conclusions

Advances in sensory methodology since the 1960s offer proven tools for margarine research and development. The importance of the various characteristics of tablespreads may now be defined through the application of multivariate statistical techniques, such as multidimensional analysis and principal component analysis, to data gathered by standard sensory methods. Valid relationships have been established between unit changes in physical or chemical measurements of margarine qualities and the sensory response, at least in the cases of hardness, spreadability, and saltiness. Such psychophysical functions permit a streamlined approach to product development provided that due recognition is given to the impact of texture-flavor interactions in the selection of sensory tests and in the training or calibration of the critical instrument, the sensory panel.

## References

1. Vaisey-Genser, M. and Carr, R.A. In *Encyclopaedia of Food Science, Food Technology and Nutrition*, edited by D. Macrae, R. Robertson and M. Sadler, Academic Press, London, 1993, pp. 2880–2884.
2. F.A.O. *Codex Standard for Margarine No. 32-1981*. Codex Alimentarius Commission, Food and Agriculture Organization of the United Nations, Rome, 1981, pp. 183–190.
3. *Consumer Reports*. 1983 *48(1)*:15.
4. Moskowitz, H.R., and Jacobs, B.E. 1986. *Acta Alimentaria 15*:29.
5. Dritschel, M.E. 1970. *Food Engineering 42(10)*:90.
6. Carr, R.A., and Vaisey-Genser, M. In *Encyclopaedia of Food Science, Food Technology and Nutrition*, edited by D. Macrae, R. Robinson, and M. Sadler, Academic Press, London, 1993, pp. 2884–2889.
7. York, R.K. 1989. *Can. Inst. Food Sci. Technol. J. 22*:441.
8. Mackie, D.A., and Elsaesser, J. 1991. *Can. Inst. Food Sci. Technol. J. 24*:264.
9. Barylko-Pikielna, N., and Bykowska, M. 1984. *Przemysl Spozywczy 38*:357.
10. Barylko-Pikielna, N., Radzanowska, J., and Kulesza, K. 1989. *Przemysl Spozywczy 43*:135.
11. Urbach, G., Stark, W., and Forss, D.A. 1972. *J. Dairy Res. 39*:35.
12. Tuorila, H., Matuszewska, I., Hellemann, U., and Lampi, A. 1989. *Food Qual. Pref. 1*:157.
13. Matuszewska, I., Barylko-Pikielna, N., Tarkkonen, L., Hellemann, U., and Tuorila, H. 1992. *Food Qual. Pref. 3*:47.
14. Vane, B.K. *Textural Assessment of Margarine with Emphasis on Canola-Based Products*. M.Sc. Thesis. Foods and Nutrition, University of Manitoba, 1982.
15. Chrysam, M. In *Bailey's Industrial Oil and Fat Products*, edited by T.H. Applewhite, Wiley and Sons, New York, 1985, pp. 41–126.
16. van Stuyvenberg, J.H. In *Margarine: an Economic Social and Scientific History 1869–1969*, edited by J.H. van Stuyvenberg, University of Toronto Press, Toronto, 1969, pp. 281–327.
17. Heick, W.H. In *A Propensity to Protect: Butter, Margarine and the Rise of Urban Culture in Canada*, Wilfred Laurier University Press, Waterloo, Ontario, 1991, pp. 134–146.
18. *Munsell Book of Color, Neighboring Hues Edition, Matte Finish Collection*. MacBeth Color and Photometry Division, Kollmorgan Corporation, Newburgh, New York, 1976.
19. *ASTM Standards on Color and Appearance Measurement*, 3rd edition, American Society of Testing and Materials, Philadelphia, 1991.
20. AOCS Official Method Cd 10-57, revised 1974, *Official and Tentative Methods of the American Oil Chemists' Society*, 3rd edition, edited by R.O. Walker, The American Oil Chemists' Society, Champaign, Illinois, 1979, pp. 1–5.

21. Hannewijk, J., Haighton, A.J., and Hendrikse, P.W. In *Analysis and Characterization of Oils, Fats, and Fat Products*, edited by H.A. Boekenoogen, Interscience Publishers, New York, 1964, pp. 119–182.
22. Weiderman, L.H. 1968. *J. Am. Oil Chem. Soc. 45:*515A.
23. deMan, L., deMan, J.M., and Blackman, B. 1989. *J. Am. Oil Chem. Soc. 66:*128.
24. deMan, L., Postmus, E., and deMan, J.M. 1990. *J. Am. Oil Chem. Soc. 67:*323.
25. deMan, J.M., Dobbs, J.E., and Sherman, P. In *Food Texture and Rheology*, edited by P. Sherman, Academic Press, London, 1979, pp. 43–54.
26. Lee, W.E. 1986. *J. Food Sci. 31:*249.
27. Chapman, D. 1962. *Chem. Rev. 62:*433.
28. Brandt, M.A., Skinner, E.Z., and Coleman, J.A. 1963. *J. Food Sci. 28:*404.
29. Vaisey-Genser, M., Vane, B.K., and Johnson, S. 1989. *J. Text. Stud. 20:*347.
30. Hough, G., Martinez, E., and Contarini, A. 1990. *J. Dairy Sci. 73:*604.
31. Haighton, A.J. 1965. *J. Am. Oil Chem. Soc. 42:*27.
32. Rohm, H., and Ulberth, F. 1989. *J. Text. Stud. 20:*409.
33. Haighton, A.J. 1959. *J. Am. Oil Chem. Soc. 36:*345.
34. Rohm, H., and Raaber, S. 1992. *J. Food Sci. 57:*647.
35. Rohm, H., and Raaber, S. 1991. *J. Sensory Studies 6:*81.
36. Riel, R.R. 1960. *J. Dairy Sci. 43:*1224.
37. *Consumer Reports* 1979. *44(2):*66.
38. Board, P.W., Aicken, K., and Kuskis, A. 1980. *J. Food Technol. 14:*277.
39. McBride, R.L., and Richardson, K.C. 1983. *Lebensm.-Wiss. u-Technol. 16:*198.
40. Dixon, B.D., and Parekh, J.V. 1979. *J. Text. Studies 10:*421.
41. Pokorný, J., Davídek, J., Dobiášová, S., Štern, P., and Čmolík, J. 1984. *J. Text. Stud. 15:*395.
42. Moskowitz, H.R. 1977. *Crit. Rev. in Food Sci. Nutr. 9:*41.
43. Christensen, C.M. In *Food Texture,* edited by H. Moskowitz, Marcel Dekker, New York, 1984, pp. 129–144.
44. Rohm, H. 1990. *Lebensm.-Wiss. u.-Technol. 23:*550.
45. Warland, R.H., and Hermann, R.O. 1971. *J. Cons. Aff. 5:*56.
46. Kleyn, D.H. 1992. *Food Technol. 46:*118.
47. Ramshaw, E.H. 1974. *Austral. J. Dairy Tech. 29:*110.
48. Lindsay, R.C., Day, E.A., and Sather, L.A. 1967. *J. Dairy Sci. 50:*25.
49. Siek, T.J., Albin, I.A., Sather, L.A., and Lindsay, R.C. 1969. *J. Food Sci. 34:*265.
50. Patton, S., and Josephson, D.V. 1957. *Food Res. 22:*316.
51. McDaniel, M., Sather, L.A., and Lindsay, R.A. 1969. *J. Food Sci. 34:*251.
52. Moskowitz, H.R. 1977. *J. Food Qual. 3:*195.
53. Fyfe, B., and Vaisey-Genser, M. 1983. *Can. Inst. Food Sci. Technol. J. 16:*206.
54. Ismail, F., and Vaisey-Genser, M. 1981. *J. Food Sci. 46:*1241.
55. Larmond, E. In *Laboratory Methods for Sensory Evaluation of Food*, Agriculture Canada Publication 1637, 1977, 77 pp.
56. Gutschmidt, J. 1974. *Lebensm. Wiss. u.-Technol. 7:*137.
57. deMan, J.M. 1969. *J. Text. Stud. 1:*109.
58. van den Tempel, M. 1958. *Rheologica Acta 1:*115.

# Appendix A

### *Laboratory Margarine Recipe*[a]

Bench-top preparation of margarine may be useful for preliminary decisions on flavoring. In these instances, a reasonable, though slightly grainy product, may be prepared using a home-type food processor with the following directions:

| | | |
|---|---|---|
| Ingredients: | Oil | 803.00 g |
| | Milk | 118.00 g |
| | Water | 78.88 g |
| | Salt | 0.12 g |

Method:

1. Chill milk-water-salt brine to 7–10°C.
2. Melt oil completely; stir in flavorants and colorants.
3. Chill the oil basestock until it is in a molten, semisolid form (at −8°C about 1 h).
4. Mix the cold oil phase in the food processor for 2–3 min with mixing blade adjusted to provide no more than 1/4" clearance from the container's inside base.
5. Add chilled milk-brine very slowly while processing at slow speed to emulsify.
6. Refrigerate to "set up."

[a]Recipe courtesy of CSP Foods, Saskatoon, Saskatchewan.

Chapter 6

# Methods to Measure Volatile Compounds and the Flavor Significance of Volatile Compounds

**R. Przybylski and N.A. Michael Eskin**

Department of Foods and Nutrition, University of Manitoba, Winnipeg MB R3T 2N2, Canada.

## Introduction

Volatile compounds are directly responsible for the flavor of fats and oils so their identification is important. A recent interlaboratory study conducted by AOCS to assess the quality of different oils showed a wide range in volatile compounds reported for the samples by participating laboratories. In light of these results, the tasks of initiating volatile analysis and building a database for use in quality control are difficult to achieve for all fats and oils. The complete volatile compound patterns formed from different fatty acids during deterioration also contributes to the difficulty. This chapter will discuss how volatile compounds can be measured by gas chromatography.

## Volatile Formation

Formation of volatile compounds from fats and oils has been investigated thoroughly in a number of reviews (1-4). This chapter will discuss volatile compound formation in relation to analyses and preparation of oil/fat databases. Many volatile components formed from the decomposition of hydroperoxides of unsaturated fatty acids in oils or fats have been identified; in some cases, a relationship has been established between their quantity and flavor properties (3). Volatiles formed during lipid oxidation were identified primarily from model systems of methyl esters of fatty acids (1).

The variety of flavor compounds produced appears to depend on the type of oxidation mechanism, the hydroperoxides formed, and the nature of the fatty acids involved. Figures 6.1–6.6 summarize the volatile components formed during oxidation of the methyl esters of oleic, linoleic, and linolenic acids via decomposition of their hydroperoxides. Two distinct oxidation mechanisms are involved, free radical and photosensitization. Each fatty acid produces different hydroperoxides depending on the number of double bonds and the oxidation mechanism involved. Further decomposition of these intermediate compounds results in the formation of a wide variety of volatiles. The schemes include all possible components formed from oleic, linoleic, and linolenic acid hydroperoxides (1). The quantitation of volatiles revealed that aldehydes accounted for almost 50%, while hydrocarbons and alcohols each contributed 14% of all volatiles formed. Thus, the total contribution by these three groups accounted for approximately 80% of all volatiles analyzed.

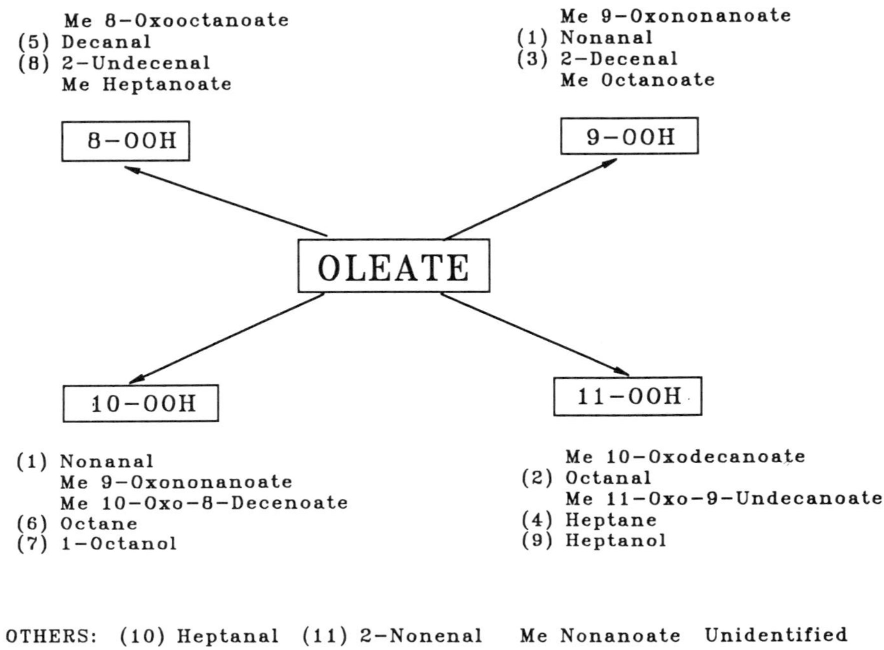


**Figure 6.1.** Formation of volatile components from the decomposition of oleate hydroperoxides (X-OOH) through free radical oxidation. Ascending numbers in brackets indicate decreasing amounts of compounds formed. The term "Others" indicates volatiles common for all hydroperoxides formed (adopted from references 5,6,52).

When fatty acid hydroperoxides are part of glycerides, volatiles such as methyl esters of short chain fatty acids and their oxidized derivatives are not produced. However, oxidized fatty acid bonded to glycerol can undergo inter- or intramolecular reactions forming rather large molecules and producing few volatiles (5). Under these circumstances the contribution of aldehydes, alcohols, and hydrocarbons account for 91% of all volatiles formed (6). Of these, aldehydes contribute approximately 60%, while alcohols and hydrocarbons each contribute 17% (see Figures 6.1–6.3). The main volatiles produced from oleate are aldehydes, such as octanal, nonanal, decanal, and 2-decenal, together with the hydrocarbons, heptane and octane (see Figures 6.1 and 6.2). The predominant volatiles for linoleate are hexanal, 2,4-decadienal, heptanal, while the hydrocarbon pentane is the most abundant (see Figures 6.3 and 6.4). With respect to linolenate (see Figures 6.5 and 6.6), the major volatiles formed are unsaturated aldehydes, 2,4,7-decatrienal, 2,4-heptadienal, propanal/propenal, and 2-butenal together with hydrocarbons, ethane/ethene (the ascending numbers in Figures 6.1–6.6, represent the possible descending order of each compound's contribution).

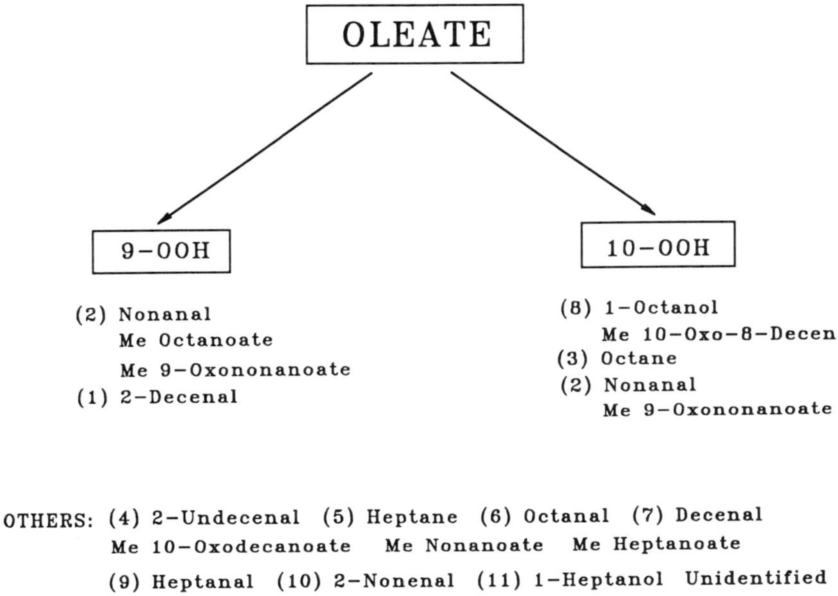


**Figure 6.2.** Volatile compound formation from oleate hydroperoxides by photosensitized oxidation. For detailed explanations see Fig. 6.1.

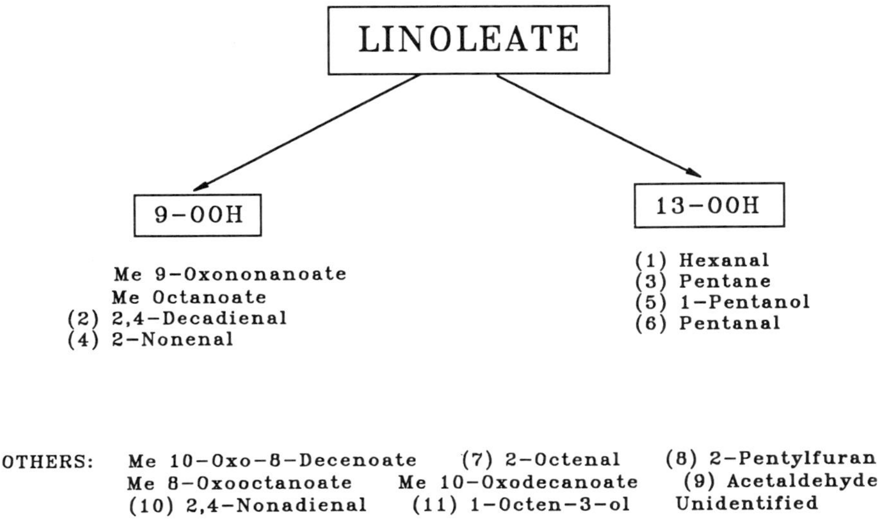


**Figure 6.3.** Flavor volatile compounds formed from the decomposition of linoleate hydroperoxides by way of free radical oxidation. For detailed explanations details see Fig. 6.1.

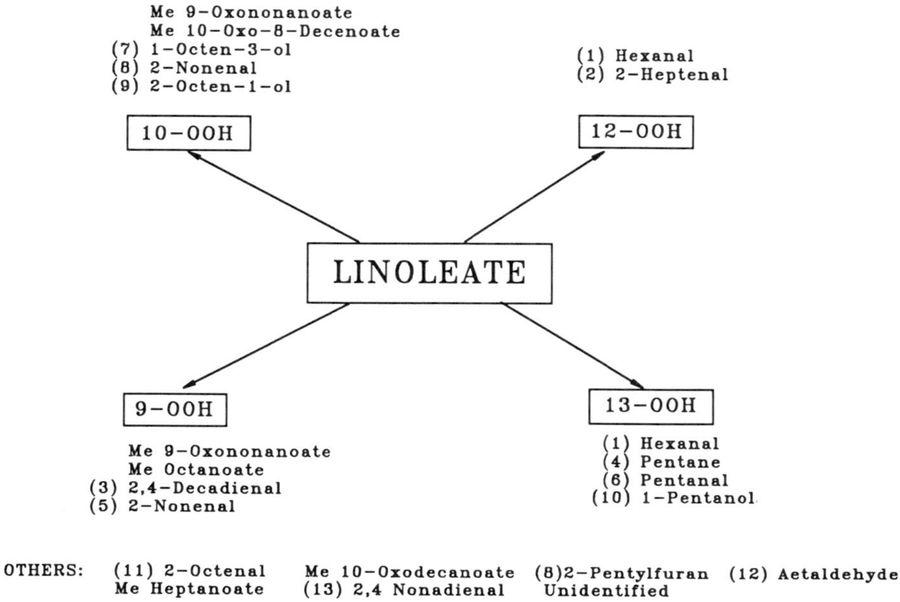


**Figure 6.4.** Volatile compound formation during photosensitized oxidation of linoleic acids. For detailed explanations see Figure 6.1.

During oxidation, cyclic hydroperoxides are produced from fatty acids with two or more double bonds. These compounds also generate another group of volatiles similar to those formed from the chain hydroperoxides (Figures 6.7 and 6.8). During glyceride oxidation, the typical volatile esters formed from fatty acid methyl esters will not be present, so that only aldehydes, alcohols, and hydrocarbons are formed. Decomposition of the cyclic hydroperoxides formed from linoleic and linolenic acids produce mainly hexanal, 2-heptenal, pentane, and ethane; or heptanal, propanal/propenal, 2,4-heptadienal, and butanal, respectively. These components are marked with numbers in brackets, ascending numbers indicate the decreasing order of possible contribution.

The pathways leading to volatile formation from hydroperoxides are incomplete without including secondary degradation to flavor compounds (7,8). Those most prone to degradation are unsaturated aldehydes which can form additional amounts of saturated aldehydes, such as hexanal and heptanal in addition with some unsaturated aldehydes. Aside from these volatiles, a variety of short-chain saturated and unsaturated fatty acids are also produced as demonstrated in Figure 6.9.

It is clear that the system involved in the production of flavor compounds from oxidized lipids is extremely fragile and dynamic. Analytical procedures used for volatile compound analysis can change the volatile profile if harsh conditions are

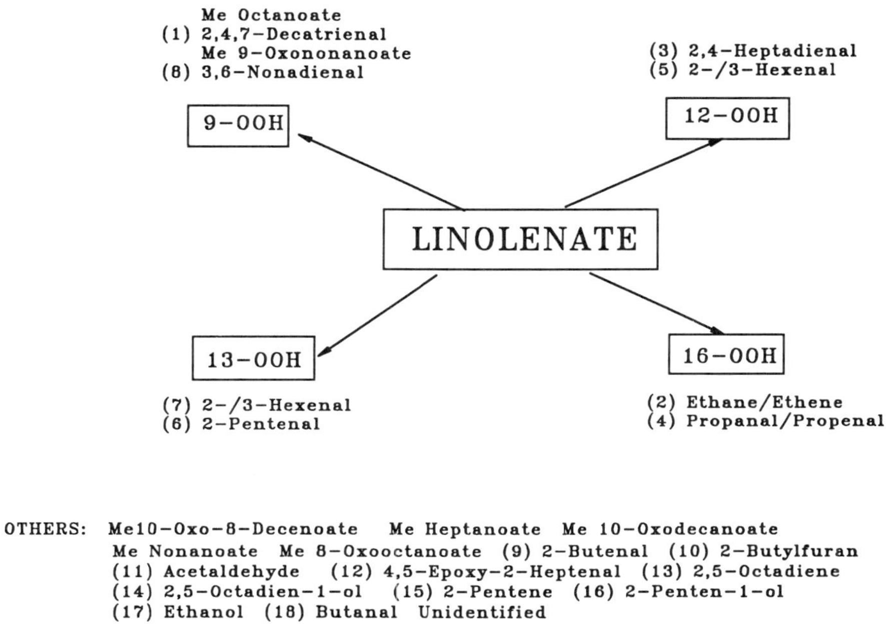


**Figure 6.5.** Flavor component formation during free radical oxidation of linolenic acid. For detailed explanations see Figure 6.1.

applied. Schieberle and Grosch (3) showed that at 38°C unsaturated aldehydes decomposed within 2 hr. Almost 40 years ago, Cooney et al. (9) reported that the stability of hydroperoxides is highly dependent on temperature and time. Figure 6.10 shows that the decomposition rate for hydroperoxides increases when temperature and time of exposure to heat is increased. At elevated temperatures this relationship was found to be nonlinear.

Vegetable oil is a far more complex system compared to model systems of fatty acid methyl esters in which interactive types of reactions can occur. In this system, reactions can be stimulated or inhibited among fatty acids and other minor components present (1,4,10). In fatty food systems, interaction of other components can occur which can stimulate the formation of a whole range of volatile compounds not directly related to lipid oxidation and deterioration (2,11–13).

## Significance of Volatiles

The myriad oxidation products formed during lipid oxidation are well documented in the literature. Gas chromatographic profiles are abundant as chemists undertake

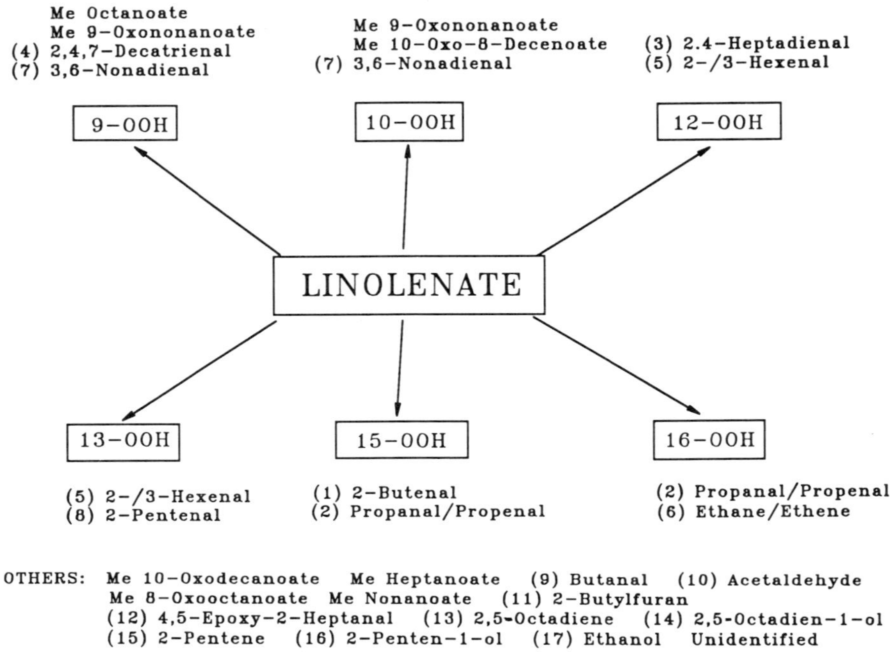


**Figure 6.6.** Formation of volatile compounds during photosensitized oxidation of linolenic acid. For detailed explanations see Figure 6.1.

the task of identifying many of these compounds (6,14–18). In contrast, there remains a lack of information regarding the contribution of many volatiles to off-odor or rancidity. In addition, it is well known that flavors can arise through the complex interaction between volatiles, and not always from a single component,

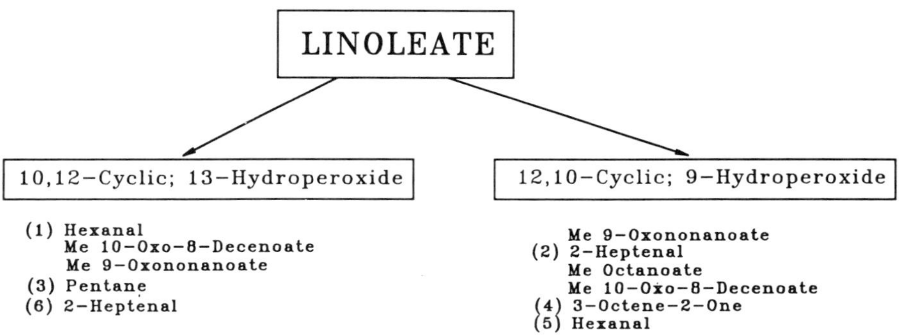


**Figure 6.7.** Volatile compound formation from the decomposition of linoleic acid cyclic hydroperoxides during photosensitized oxidation. For detailed explanations see Figure 6.1.

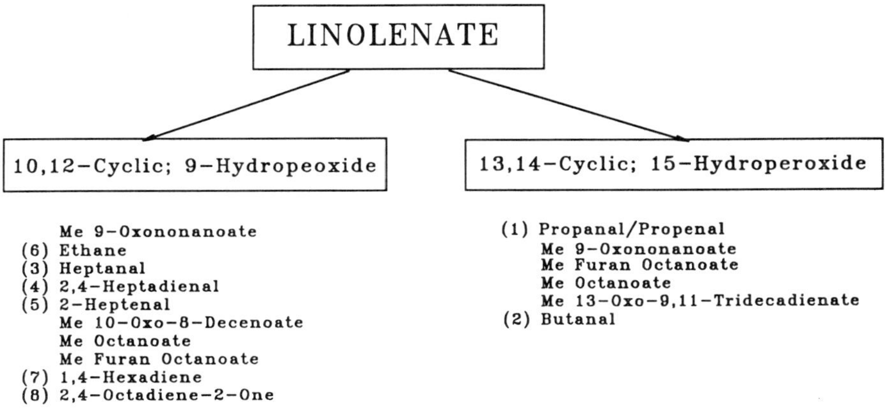


**Figure 6.8.** Formation of flavor compounds from the decomposition of cyclic hydroperoxides of linolenic acid formed during photosensitized oxidation. For detailed explanations see Figure 6.1.

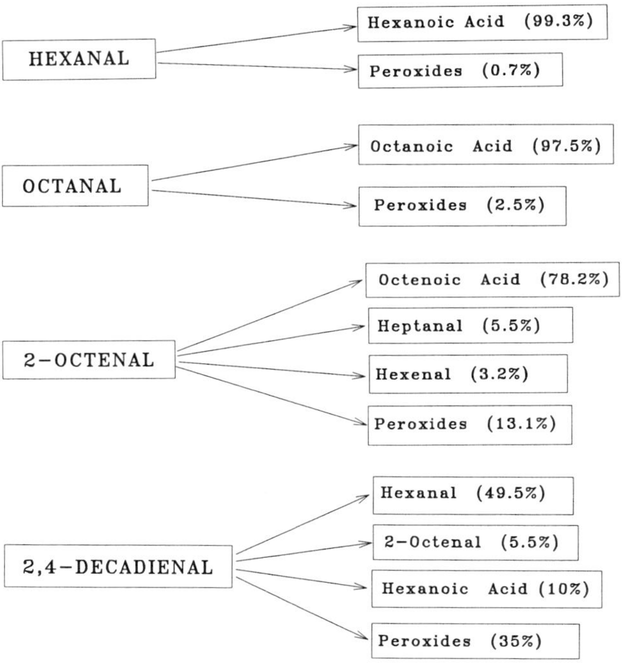


**Figure 6.9.** Secondary decomposition of the volatile components formed from fatty acid oxidation. Percentages in parentheses are the concentration for one volatile compound.

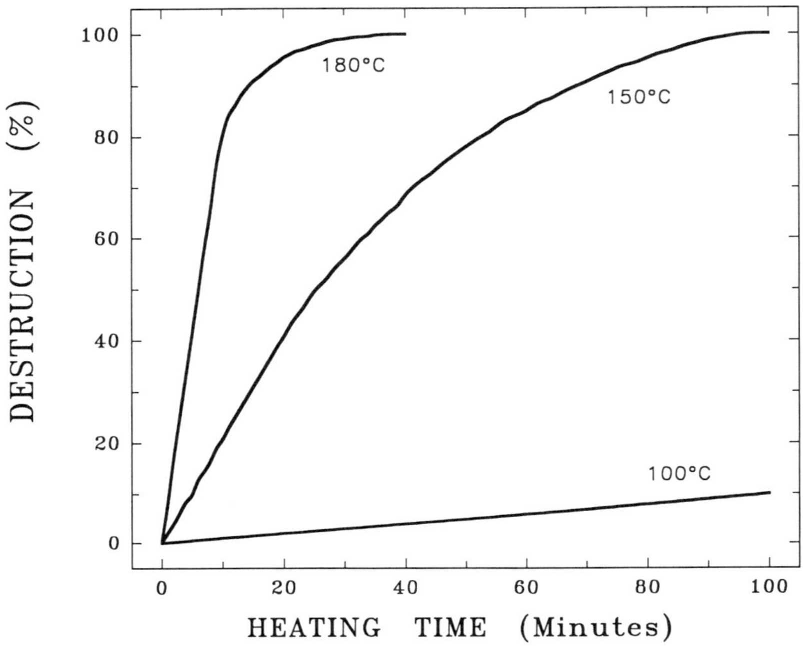


**Figure 6.10.** Effect of temperature and time on the stability of fatty acid hydroperoxides (adopted from Cooney et al. [9]).

making the possibilities almost infinite in number. This is further complicated by the confusion resulting from different terms used by panelists to describe many of these flavors. In spite of these difficulties, once a component has been identified, several characteristics must be established, including individual threshold values, flavor character at different concentrations, and the amount present in the sample under different oxidation conditions.

Such information is extremely important since a certain component may be present at very low concentrations but be exceptionally influential in terms of its contribution to flavor. For example, the potent volatile identified in soybean oil appeared to be 1-octen-3-one (19). When calculated on a weighted percentage, *trans, cis*-2,4-decadienal was found to be the most significant flavor volatile followed by 1-octen-3-one, *n*-butanal, and *n*-hexanal (Table 6.1). Hydrocarbons, such as pentane, generally exhibit the highest threshold values and are thus insignificant with respect to their contribution to off-odor or rancidity.

**TABLE 6.1**
Flavor Significance of Soybean Oil Volatiles[a]

| Major volatiles | Relative % | Threshold value[b] (ppm) | Weighted % (1-Octen-3-ol) | Relative order |
|---|---|---|---|---|
| *t,t*-2,4-Decadienal[c] | 33.7 | 0.10 | 2.5 | 2 |
| *t,c*-2,4-Decadienal[c] | 17.9 | 0.02 | 6.7 | 1 |
| *t,c*-2,4-Heptadienal[c] | 11.1 | 0.04 | 2.1 | 3 |
| 2-Heptenal | 5.6 | 0.2 | 0.21 | 8 |
| *t,t*-2,4-Heptadienal[c] | 4.5 | 0.1 | 0.34 | 7 |
| n-Hexanal | 4.5 | 0.08 | 0.42 | 6 |
| n-Pentane | 3.1 | 340 | $6.8 \times 10^{-5}$ | 16 |
| n-Butanal | 1.5 | 0.025 | 0.45 | 5 |
| 2-Pentenal | 1.2 | 1.0 | 0.009 | 13 |
| 1-Octen-3-ol | 0.9 | 0.0075 | 0.9 | 4 |
| 2-Pentyl furan | 0.8 | 2.0 | 0.003 | 14 |
| n-Pentanal | 0.7 | 0.07 | 0.075 | 10 |
| 2-Hexenal | 0.7 | 0.6 | 0.009 | 13 |
| n-Nonanal | 0.7 | 0.2 | 0.026 | 11 |
| n-Heptanal | 0.6 | 0.055 | 0.082 | 9 |
| 1-Penten-3-ol | 0.5 | 4.2 | $8.9 \times 10^{-4}$ | 15 |
| 2-Octenal | 0.5 | 0.15 | 0.025 | 12 |

[a] Data from reference (67) (Room temperature air. peroxide value 9.5).
[b] Data from reference (68).
[c] *t,t* = *trans,trans*; *t,c* = *trans,cis*.

# Methods Used to Measure Volatiles

## *Characterization of Samples*

Analysis of volatile components in oil and fat-bearing foods is a difficult task. Most volatiles are very soluble in lipids and are usually present at very low concentrations. The main problems related to analysis of flavor compounds in these samples are the oil viscosity, oil affinity for volatiles, contamination, and interference by additives or solvents. The high viscosity of the oil/fat at room and elevated temperatures can lower the rate of diffusion and transfer of volatiles to the analytical system. The high affinity of the oil/fat for volatiles can affect recovery and measurement, because of the superior solubility of flavor compounds in these solvents (21). In addition this can directly influence the evaluation of their flavor significance, as the threshold level is usually higher for this same compound in oil than in water (22). Artifacts may be present in the oil or fat due to contamination from the storage area and laboratory organic chemicals. This can cause significant problems associated with identification and assessment of compounds not normally present or formed in the oil or fat. Both oil additives and solvents can produce additional peaks on a chromatogram and affect flavor, making the analysis more difficult to perform (23). Oil chemists face additional problems when analyzing volatiles in fatty food products, such as potato chips, mayonnaise, and french fries, because of moisture, nonlipid components, and a solid or semisolid form.

Moisture, usually present at levels ranging from 2–90%, can improve volatile transfer (24,25). Simultaneously, it can limit the absorption capacity of sorbents used or interfere with the analytical system, for example, by blocking the system with ice formation during the cryogenic collection of volatiles.

Components other than lipids present in foods can interact during the processing and analytical procedure and form additional volatiles not directly related to oil or fat deterioration. In addition, lipid-oxidation products can also react with other food components. For example, aldehydes and proteins generate a variety of new flavor compounds (26–28). This complex mixture of volatiles requires highly efficient columns to be able to separate hundreds of volatiles. Pre-separation is required to isolate the fraction of chemical groups of flavor components present (29).

Most foods in a solid or semisolid form present similar problems to an oil or fat system. For instance, a longer equilibration time is required to produce an adequate volatile concentration in the headspace to more accurately reflect their presence in the product. The rate of diffusion is lower in solid matter than in a liquid (21).

## Analytical Methods

The most widely used methods for volatile analysis are static headspace, dynamic headspace, and direct gas chromatography (GC). A number of excellent reviews have been published using these methods to analyze volatiles in oils, fats, and fatty foods (30–32). The following discussion will cover the basic principles of these methods with their advantages and limitations.

Of the methods used for volatile analysis in oil, those with the simplest procedures and lowest sample temperatures should be selected. Of the methods presented, direct gas chromatography is probably one of the best methods to choose. Because it is difficult to automate, this technique requires a skilled operator and may be too labor intensive for some laboratories.

### *Static Headspace*

One of the first methods for analyzing volatiles was the classical headspace method (33). This very simple method is based on transferring a certain volume of equilibrated gases from the area above a lipid sample in a closed container. This gaseous sample is injected directly into the column of a gas chromatograph where separation and quantitation is performed. As mentioned previously, viscosity and solubility mainly affect the time and temperature required to obtain equilibrium between the lipid sample and gaseous headspace above it. Based on the theory of headspace, proper sampling can only be performed when equilibration is achieved, and may take up to several hours in lipid samples.

One of the major problems associated with early work on headsp necessity to maintain the sample at elevated temperatures to achieve equ (21,33). These conditions can alter the volatiles present in the lipid sample as a result of decomposition, interaction, and oxidation reactions. Differences in solubility and partial pressure of individual components can also affect their transfer, so that the volatiles in the headspace may not accurately reflect the composition of these same components in the lipid sample (33).

*Procedure.* Static headspace can be performed as a single injection from a sample where equilibrium is readily achieved. This is true for most aqueous samples, but is harder to accomplish for lipids and fatty foods. With respect to the more difficult samples, much better results can be achieved using multiple headspace extraction (41). In this case, multiple injections are performed on a single sample to extract as much of the volatile compounds as possible (18). This approach is less affected by nonequilibrated samples, but multiple injections require more time to perform volatile analysis (18,41).

The availability of automated static headspace systems has simplified the analysis of oil and fat volatiles. A major problem with older systems was controlling the sample temperature. In some units, the samples usually were kept at 180°C until they were analyzed. For example, the first samples in the tray were held for 1 hr while the last samples could be held in excess of 6 hrs. These analytical systems were originally designed to analyze volatile pollutants in environmental samples exhibiting much higher stability than lipids or food systems, and were adapted for lipid and fatty food volatile analysis later. Newer systems do not heat samples until injection.

*Advantages.* This method is convenient, rapid, and reproducible when parameters are established for the analysis of a particular type of sample. High reproducibility was attained when the manual syringe injection was replaced by valve sampling with constant volume loop or an automated sampler with pressurization. However the injected volume is affected by sample pressure and temperature, necessitating the use of correction factors.

The volatile components analyzed should represent the same volatile compounds detected by sensory evaluation. This is true only when samples are heated to this same temperature in both types of analyses. In reality, the flame-ionization detector commonly used with the GC, is less sensitive than our nose and requires a much higher temperature to obtain a chromatogram. Equilibrium of volatile compounds in the headspace is temperature dependent (33).

Static headspace is a clean method for analyzing volatiles. Because only the gaseous part of the sample is evaluated, the analytical system is free from contamination and any artifacts formed. This is particularly important when oxidized samples are analyzed because additional volatiles may be formed from the decomposition of oxidation products if they have direct contact with the hot parts of the GC injector. Static headspace can be easily automated and is relatively simple to standardize for a particular sample type. Standardization is not universal for all samples however, as the concentration of analytes is affected by time, temperature, vapor pressure, and the number of components present (33). Analysis of volatile compounds by static headspace is sensitive to high and medium volatile compounds (see Figure 7.6). The sensitivity of this method is directly affected by the boiling point and vapor pressure of the compounds analyzed (33).

*Disadvantages.* The major limitation of this procedure is the importance of achieving equilibrium to analyze all volatile compounds present which is time and temperature dependent. Equilibrium is also affected by the state of the sample (solid, semisolid, or semiviscous liquid) as it is particularly difficult to reach equilibrium due to the slower diffusion rate in viscous systems when compared to gaseous systems (34). This is directly related to the temperature and amount of time required to achieve equilibrium (33). Both temperature and time should be controlled with great accuracy in order to achieve reproducible GC results (33).

Elevated temperatures are required to analyze any high boiling point components present whenever trace component analysis is involved. To analyze volatile compounds by static headspace, oil samples are often heated to 160–200°C for 10–20 min (35–38). This can alter the volatile composition of the sample by forming additional volatile compounds from hydroperoxides and lipids (3,9,39,40).

There are also a number of technical problems associated with static headspace, including volatile adsorption on the seal used to close the container. Rubber and high-porosity seal materials can reduce the concentration of more polar components, such as aldehydes, and acids. Second, condensation can occur in many parts of the sample-transferring equipment (syringes, valves, loops, and transfer lines) which are colder than the sample and result in carryover and system contamination. To prevent this, the temperature of all parts in contact with the sample should be 5–10°C higher than the sample. Third, when large volumes of gaseous samples are injected in capillary columns, cryogenic focusing is indispensable for preventing peak widening and loss of separation efficiency by the column.

## Dynamic Headspace

These methods utilize a sorbent to collect volatiles from oil, fat or foods which are later transferred into a gas chromatograph for separation and quantitation (42). Volatiles are swept from the heated sample with nitrogen and trapped on the sorbent, usually held at room temperature. Transferring volatiles from the samples to the adsorbent may take up to several hours and is affected by the concentration of flavor components present. The most widely used porous sorbents are Tenax GC, Chromosorbs, and Porapak Q (43–45). Volatiles trapped on the adsorbent are then transferred into a GC column by thermal desorption or solvent extraction/desorption, both of these techniques will be discussed later.

*Advantages.* This method does not require the sample to be in equilibrium with the headspace. In order to be quantitative, the sweeping gas constantly removes volatiles from the sample container providing a continuous extraction. The detection limit for this method is in the range of ppt, so that components with low volatility can be analyzed accurately. This allows the analysis of a wide range of compounds, quantitatively and with high reproducibility.

Utilization of gas extraction allows a lower temperature to be used during purging, preventing decomposition of volatiles or their precursors. During purging, a high recovery is usually achieved because there is a large contact area between sample and the flowing gas. This results in a continuous extraction, which stimulates volatile diffusion from the sample interior to the sample surface (46).

Different adsorbents can be applied during single trapping to prefractionate volatiles. The adsorption characteristics determines which sorbent is used (47).

*Disadvantages.* Collecting and concentrating volatiles using adsorbents has a number of drawbacks limiting their application and requiring additional cautionary measures during operation. Each porous sorbent has its own characteristics and properties, such as area, chemical composition, breakthrough volume (volume of gas required to move particular components through the plug), and gas flow rate, all of which determine the type and amount of compounds retained. Adsorbents such as Tenax and Chromosorbs adsorb components with medium and high boiling points, while Porapak adsorbs more volatile compounds (45,48).

Dynamic headspace requires intensive manipulation when transferring volatiles from the sample to adsorbent. Water must be removed from the adsorbent by purging with dry gas, since the smallest amount can interfere with the analysis. The final manipulation involves transferring volatiles from the adsorbent into the GC system for analysis. Each of these steps can result in the alteration of compounds and formation of artifacts. This can occur if oxygen is allowed to come into contact with volatiles and the sorbent (49,50). This is particularly important during thermal desorption, which is usually performed at 150–200°C, and could stimulate oxidation of both analyzed volatiles and sorbents. Multistep manipulation with volatiles also requires strict control of the parameters to ensure reproducibility. All these complicated operations require costly equipment.

The volatiles transferred are very sensitive to cold spots in the system where condensation can take place, causing losses of some compounds. Tubing, connections, and valves all have to be heated to avoid condensation. The presence of certain metals can also catalyze decomposition and alteration of volatiles. Some of the containers used for sampling are difficult to keep clean. This can result in the carryover of volatiles between samples and oxidation can be stimulated by a thin layer of oxidized oil or fat remaining on the walls of the container.

Control of the quantity of volatiles transferred to the GC column is difficult. Overloading is most likely to occur in capillary columns.

## Thermal Desorption

A critical step in dynamic headspace is the desorption method used to release the volatiles. Thermal desorption is the most common method and is carried out at elevated temperatures using a flow of purging gas opposite to the flow involved in collecting the volatiles.

*Advantages.* Volatiles are transferred directly from the sorbent to the GC column, eliminating dilution because analytes are condensed in the column using cryogenic focusing. This form of desorption is very effective and increases the detection limit of compounds separated on capillary columns. Thermal desorption avoids the use of solvents to recover volatiles, making it easier to analyze highly volatile compounds. Not only does desorption transfer volatiles from the sorbent to the GC column, it also cleans the sorbent so that it is ready to be reused. This shortens the time required for volatile analysis.

*Disadvantages.* High temperature is required to release volatiles from the sorbent, usually in the range of 150–200°C for 5–30 min. This can cause decomposition of unsaturated aldehydes, the main oxidation products formed from lipids, as well as possible interaction between components (6). Contamination with oxygen during manipulation will diminish the amount of aldehydes, because they will be oxidized to short-chain fatty acids and other products (3,7). Elevated temperatures also affect the integrity of polymers employed. These organic products slowly decompose when heated to the elevated temperatures required for desorption, with any artifacts formed contaminating sample volatiles (40,48). Multidesorption is often practiced, increasing the possibility of contamination with artifacts from sorbent.

Cryofocusing (cryogenic gas chromatography) can promote aerosol formation, particularly in open tubings. Aerosol formation will cause sample losses during desorption.

## Solvent Desorption

The use of solvents for volatile desorption has its own limitations. Desorption of volatiles with solvents is performed at ambient temperatures to prevent heat sensitive compounds from decomposing. The most common solvents used for desorption are light hydrocarbons, such as pentane and diethyl ether, a volatile and polar solvent. Carbon disulfide, an inorganic solvent not detected by the flame-ionization detector, is used also, although the volumes normally injected can still cause baseline disturbances.

*Advantages.* Several injections can be performed. Because thermal desorption can cause decomposition and incomplete recovery for some compounds, solvents result in better recovery (47). Sample recovery from sorbent is less complicated and easier to perform because fewer parameters need to be controlled. Solvent extraction is less expensive and does not require any complicated equipment for desorption or cryogenic attachment for the gas chromatograph.

*Disadvantages.* The presence of highly volatile solvents may mask some compounds with similar retention times. A minimum volume of solvent is required to extract analytes from the sorbent. This often dilutes the volatile compounds present at very low concentrations and results in nondetection due to the lower detection limit. Evaporation of excess solvent to concentrate analytes can be a major source of error in quantitation because some compounds can evaporate with the solvent.

The use of solvents introduces possible artifacts or contamination from solvents, glassware, or storage containers. In addition, the solubility of all compounds is not the same in any solvent and will determine desorption efficiency. Solvent desorption is also limited by sorbent characteristics, its solubility, and potential contamination of the analytical system. Sorbents often have to be extracted for several hours with the solvent before it can be used for volatile collection.

### *Direct Gas Chromatography*

This method was developed for analyzing volatiles in oils, fats, and fatty foods. Three different versions of this methodology are used in many laboratories. One version uses an external purging system (24,51). Volatiles are transferred from the sample and placed in an external unit with temperature control before being injected directly into the GC column, usually packed with Tenax coated with 7% Poly MPE (51). Another version of this technique uses the same column packing, but the oil, fat, or other product is placed in an injector insert positioned in the injection port of a GC. Volatiles are transferred directly to the packed column at room temperature or using cryogenic chromatography on a capillary column (16,52,53). A third version of direct chromatography is a modification of previous methods where the chromatographic set-up is similar, except that the sample of oil or fat is injected directly into an injection port where the insert with glass wool plug is placed. When using a packed column with sorbent, volatiles are trapped at ambient temperatures, but with capillary columns cryogenic trapping/focusing is recommended (54,55).

*Advantages.* Direct gas chromatography has most of the benefits of purge-and-trap, including a high detection limit and higher enrichment. The procedure is simpler than dynamic headspace, because the direct transfer of volatiles from the sample to the column provides better recovery and quantitation. Little data is available describing the efficiency of direct gas chromatography (24,52). Using this technique, volatiles from methane to dodecadienal and its isomers can be separated in a single run (36,56).

For improved separation, capillary columns can be used which require cryogenic focusing to prevent peak broadening. Column sorbents are not needed, eliminating artifact formation. The relative simplicity of this procedure makes it inexpensive and less labor and time intensive, requiring only minor modifications to the GC.

*Disadvantages.* Chromatograms are strongly dependent on sampling conditions. Of particular importance is the injector temperature used for purging volatiles from the sample, usually around 150–190°C (36,55). Under these conditions hydroperoxides readily decompose to form additional volatiles.

When external units are used for transferring volatiles, all the disadvantages associated with static and dynamic headspace techniques apply. The use of capillary columns and cryogenic focusing can be expensive, because of the large volumes of liquid nitrogen or solid carbon dioxide needed. Direct gas chromatography is also difficult to automate.

## Volatile Compounds

### *Pentane*

One of the first volatiles used to assess lipid oxidation in oils was pentane. It was reported to be the predominant short-chain hydrocarbon formed from the thermal decomposition of oils. Jarvi and co-workers (57) monitored pentane formation as an index of rancidity in soybean oil. Fresh and abused soybean oil (aged at 60°C for 24 hr) with added *n*-octanol as internal standard, was examined by direct injection GC with a packed column. While it was not possible to separate pentane cleanly, the area under the pentane peak was measured, and the ratio of area (A) to height (h) of *n*-octanol peak in Figure 6.11 was used to describe the so-called "Oxidative Value" (OV). These researchers found an excellent correlation between OV, peroxide value (PV), and flavor.

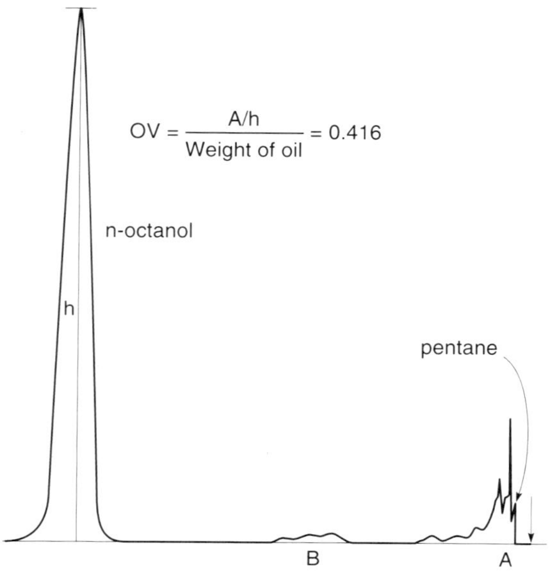


**Figure 6.11.** Gas chromatogram of soybean oil, aged 24 hr at 60°C. Oil injected directly into the column (57).

Using GC, pentane formation was later detected in the headspace of solid foods and correlated with the quality of both aged oils and potato chips. Using headspace analysis, Warner and co-workers (58) monitored pentane formation in oil and chip samples stored in jars fitted with rubber serum caps to permit sampling. All of the samples were stored at 60°C and assessed for rancid flavor. Their results showed an excellent correlation between pentane formation and flavor evaluation and a relationship between linoleate content and pentane formation. A later study by Min and Schweizer (14) also monitored lipid oxidation in potato chips stored for 9 d at 55°C against the amount of tert-butylhydroquinone (TBHQ) in the frying oil. An inverse relationship was found (Figure 6.12). These researchers also reported high correlation coefficients, suggesting that volatile changes in potato chips were the direct result of lipid oxidation.

Waltking and co-workers (15) described a relatively simple procedure for defining the quality of vegetable oils by direct injection of stored corn oil samples, spiked with an internal standard *n*-octadecane. An excellent fit was found between pentane and PV (Figure 6.13). Morrison et al. (59) attempted to correlate GC volatiles with flavor intensity for stored oils using the direct gas chromatography developed by Dupuy et al. (16,60). Using this procedure they attempted to correlate pentane and total volatiles with flavor intensity scores in sunflower oil. Their results showed that pentane increased from 0.13 to 1.18 ppm over 16 days' storage while the total volatiles ranged from 0.76–6.95 ppm. Excellent correlations were reported between flavor intensity and pentane.

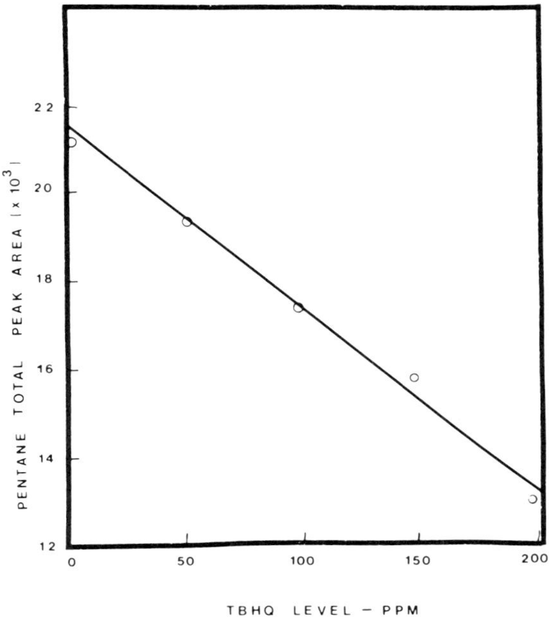


**Figure 6.12.** Pentane content in the headspace of potato chips produced in oils containing different levels of TBHQ. The samples were stored for 9 days at 55°C (14).

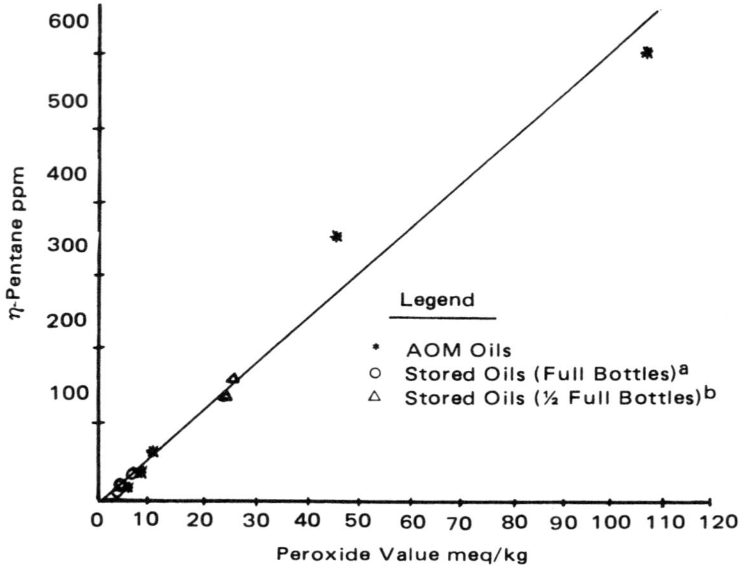


**Figure 6.13.** The relationship of pentane, peroxide value and flavor of corn oils. Flavor score based on 5 = excellent and 1 = terrible. (a) Flavor scores in the order of increasing PV (3.45, 3.70, 3.90); (b) Flavor scores in the order of increasing PV (2.35, 3.50)(15).

## *Hexanal and Propanal*

Hexanal was first detected by Buttery and Teranishi (61) in the vapors of dehydrated potatoes, which suggested it was an indicator of linoleate oxidation. Subsequent work by Fritsch and Gale (62) used hexanal for monitoring rancidity of low-fat foods, such as cereals. The food sample (15 g) was mixed with an internal standard, 4-heptanone (b.p. 144°C) and boiling water (150 mL). After swirling the mixture for 45 sec, 5 $cm^3$ of headspace gas was withdrawn and analyzed by GC. Figure 6.14 illustrates the chromatogram obtained for the stored cereal sample in which peak 8 was identified as hexanal. An increase in hexanal was observed for a number of cereal products, including oats, corn, and wheat, after being stored for 12 weeks at 37°C. Confirmation of hexanal as a lipid-oxidation product was shown by a decrease in hexanal production in the presence of 40 ppm of an antioxidant. The most effective antioxidant examined was TBHQ.

Using the direct GC method of Dupuy et al. (15), Warner and associates (63) found that hexanal and pentanal could be used to predict flavor scores for soybean, cottonseed, and peanut oils. The correlation coefficients were highest for soybean oil (–0.95 and –0.90) followed by cottonseed oil (–0.60 and –0.85) and peanut oil (–0.74 and –0.75).

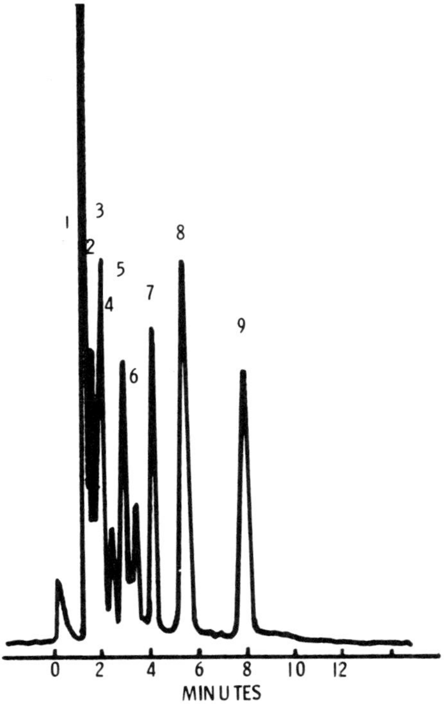


**Figure 6.14.** Chromatrogram of a stored cereal sample. Peak 9 is the internal standard 4-heptanone, peak 8 is n-hexanal and peak 7 is iso-butyl ketone which previously was used as anm internal standard (62).

A more recent study by Robards and co-workers (64) monitored the formation of propanal, pentanal, and hexanal by headspace analysis to assess rancidity of a 50:50 blend of cottonseed and sunflower oil used to fry corn chips. A typical chromatogram obtained from the headspace gas analysis of the used oil stored for 20 d under Schaal Oven conditions is shown in Figure 6.15. Among the products identified were propanal, *n*-pentanal, *n*-hexanal, 2-decenal, and minor volatile products. Pentane was not detected, although it co-eluted with propanal and might still be present in trace amounts. Of the volatiles measured, hexanal was produced in the largest quantities.

## 2,4-Decadienal

Assessment of vegetable oil quality by Dupuy et al. (16) using direct gas chromatography monitored the production of pentane, total volatiles, and *trans, trans*-2,4-decadienal. Their results showed that 2,4-decadienal was the most sensitive volatile marker with respect to its correlation with flavor scores (Figure 6.16). Min (10) monitored the formation of decadienals in stored soybean, corn, and hydrogenated cottonseed oils and found excellent correlation coefficients between volatile measurement and flavor assessment.

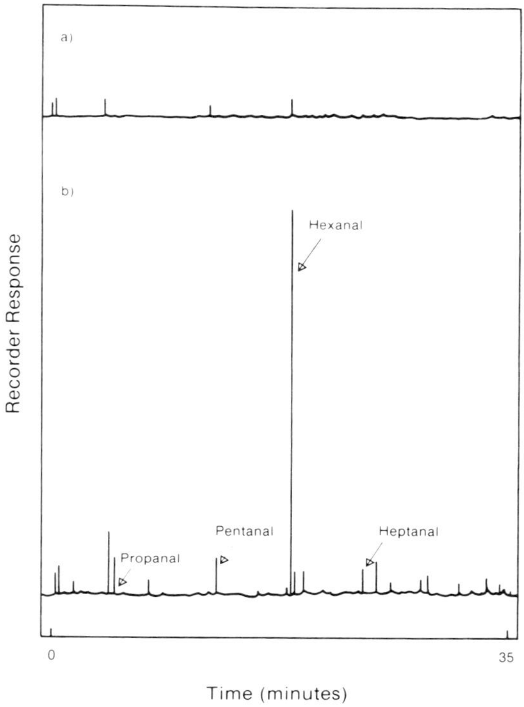


**Figure 6.15.** Chromatograms of headspace vapors from corn chips fried in a 50:50 blend of cottonseed and sunflower oils. In a, samples were analyzed one hr. after frying and in b, after storage for 20 days in the dark at 60°C. Conditions: 25 m SE-54 WCOT column programmed at 10°C/min. from 40°C to 240°C (Detection sensitivity is 10 times greater in a) (64).

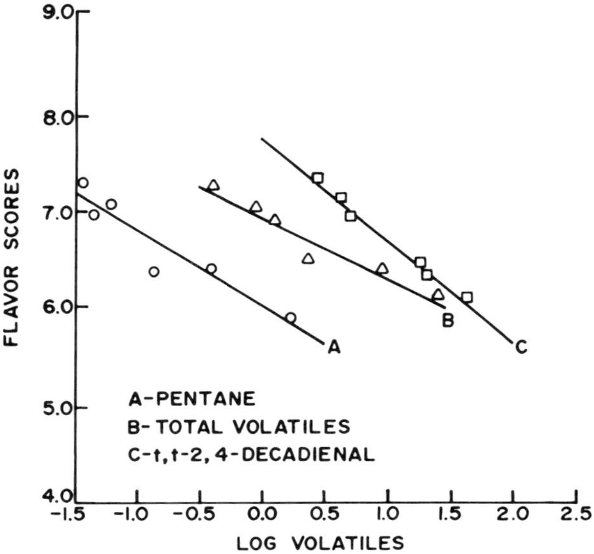


**Figure 6.16.** Regression of flavor scores on the log of the gas chromatographic volatile peaks: A, pentane; B, total volatiles; C, *trans*-2, *trans*-4-decadienal (16).

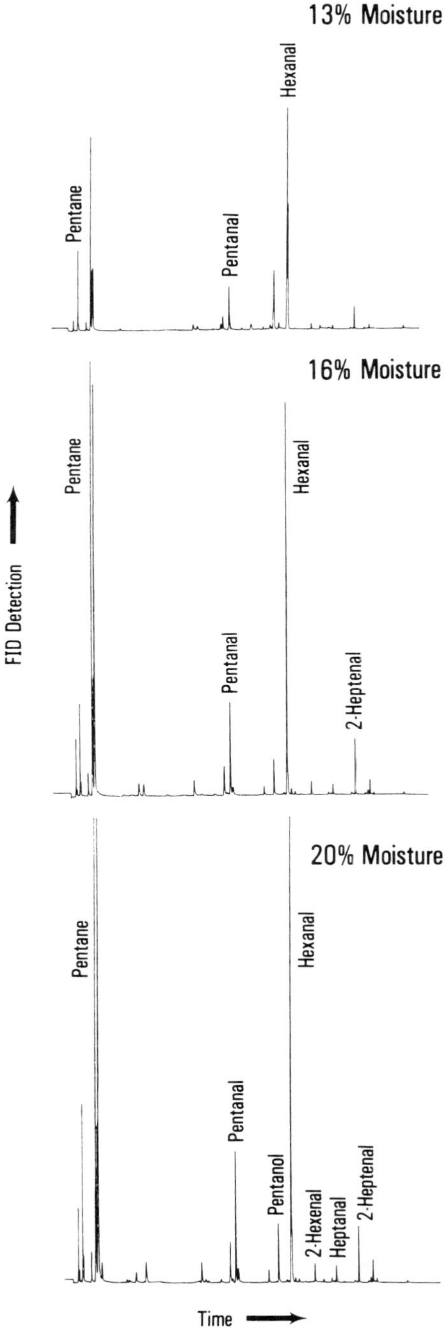


**Figure 6.17.** Capillary headspace gas chromatograms of volatiles in crude oils from soybeans stored at 13% and 16% moisture for 22 days and at 20% moisture for 21 days (37).

### *Mixed Volatiles*

The quality of soybean oil extracted from seed stored at different moisture contents was monitored by Frankel and co-workers (37) using static headspace GC. Figure 6.17 (page 127) shows the volatiles identified in the crude oils extracted from soybeans stored at 13, 16, and 20% moisture contents. The peaks identified were pentane, pentanal, hexanal, 2-hexenal, and 2-heptenal; typical oxidation products indicating a higher degree of oxidative deterioration with an increase in moisture content.

Snyder and co-workers (42) compared volatiles formed from soybean oil with those formed from the corresponding hydrogenated oils. A main volatile formed from linolenate present in this oil was 2,4-heptadienal. After reduction of the linolenate content to <0.5% by hydrogenation, only trace amounts of this compound were detected. Hexanal and 2-heptenal were present as they were formed from linoleate. A linear relationship also was observed between nonanal and oleate. Other volatiles, including 2-decenal, 3-undecenal, octanal, and nonanal, increased after hydrogenation. Raghavan et al. (42) reported a rapid method for assessing vegetable oil quality by dynamic headspace GC. The GC profiles obtained for soybean and corn oil are shown in Figures 6.18 and 6.19 with the volatiles, 2-*trans*-heptenal, 2-*trans*-octenal, and 2,4-decadienal selected for correlation with flavor scores. A

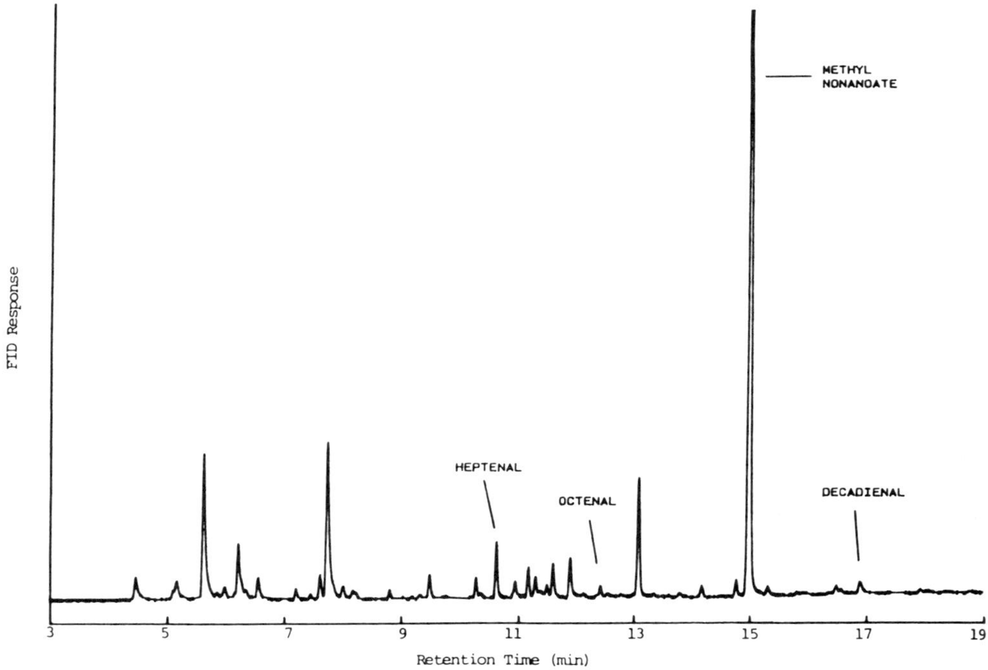


**Figure 6.18.** Profile of soybean oil volatiles purged at 150°C for 20 minutes (Methyl nonanoate added as an internal standard.) (42).

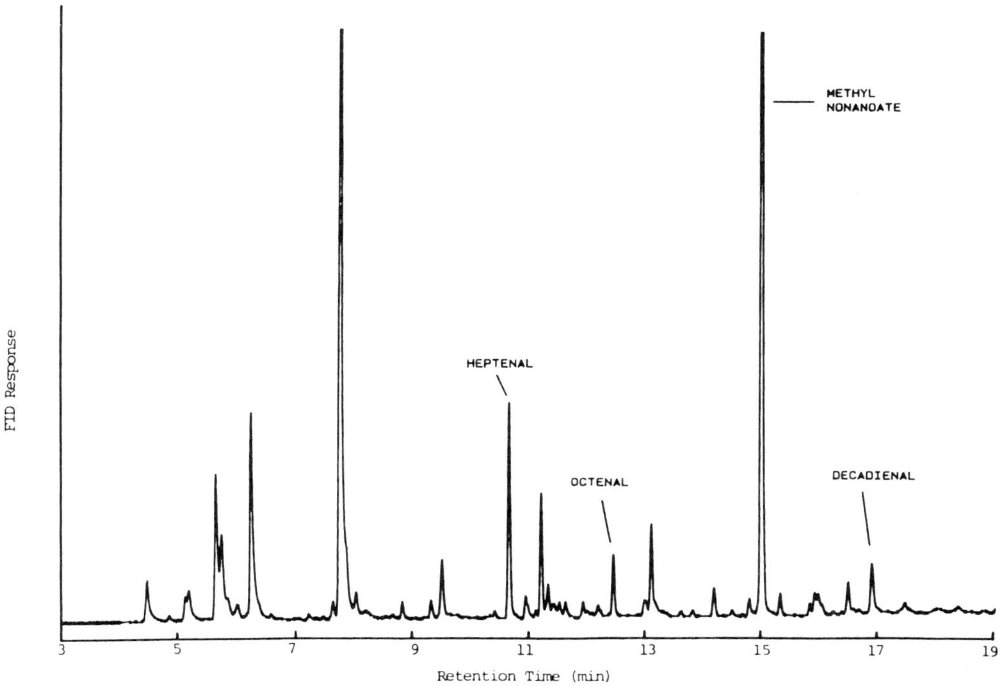


**Figure 6.19.** Profile of corn oil volatiles purged at 150°C for 20 minutes (Methyl nonanoate added as an internal standard.) (42).

good correlation was obtained between these volatiles and sensory evaluation as shown for soybean oil in Table 6.2.

A method for analyzing oil volatiles using multiple headspace extraction (MHE) was adapted by Snyder and Mounts (18). This method involves the repeated withdrawal of volatile compounds in the headspace of a solution and is equivalent to a liquid-liquid extraction using a series of separatory funnels. Twelve volatiles found in vegetable oils were identified, propanal, pentane, pentenal, pentanol, hexanal, 2-pentenol, 3-hexenal, 2-heptenal, octen-3-ol, 2,4-heptadienal, nonanal, and 2,4-decadienal. Using this technique, these researchers monitored the storage stability of corn, canola, soybean, sunflower, and high-oleic oils over an 8-day period using the Schaal Oven Test and MHE. All volatiles were shown to increase during storage except for pentanal and 1-octen-3-ol in canola oil. Pentane exhibited the largest increase in all oils, except for the high-oleic acid sunflower oil in which hexanal showed the largest increase. These researchers compared MHE with the single headspace technique (SHT) using the same oils; while some volatile changes were similar, overall MHE showed greater reproducibility than SHT and had a much lower coefficient of variation for all volatiles evaluated except pentane and propanal.

**TABLE 6.2**
Concentration of Heptenal, Octenal and Decadienal in Twelve Randomly Selected Soybean Oils with the Instrumental and Sensory Flavor Scores

| | Concentration (ppb) | | | Flavor score | |
|---|---|---|---|---|---|
| Sample | Heptenal | Octenal | Decadienal | Sensory | G.C. ($\sigma$) |
| 1 | 92 | 40 | 94 | 6.7 | 6.8 (± 0.2) |
| 2 | 83 | 18 | 65 | 6.8 | 7.0 (± 0.2) |
| 3 | 124 | 30 | 91 | 6.9 | 6.7 (± 0.1) |
| 4 | 184 | 48 | 126 | 6.9 | 6.2 (± 0.1) |
| 5 | 87 | 19 | 52 | 6.9 | 7.0 (± 0.3) |
| 6 | 207 | 25 | 93 | 6.9 | 6.6 (± 0.2) |
| 7 | 224 | 21 | 108 | 6.4 | 6.6 (± 0.2) |
| 8 | 217 | 74 | 157 | 6.5 | 5.9 (± 0.1) |
| 9 | 185 | 34 | 206 | 6.6 | 6.2 (± 0.1) |
| 10 | 75 | 10 | 69 | 6.7 | 7.0 (± 0.3) |
| 11 | 96 | 14 | 113 | 6.7 | 7.0 (± 0.2) |
| 12 | 88 | 22 | 132 | 6.8 | 6.9 (± 0.2) |
| $\bar{x}$ | | | | 6.73 | 6.66 |

## Colorimetric Methods

Colorimetric methods are often used, although gas chromatography provides more information about individual components. A complex picture of a few dozen peaks does not solve the problem of oil- or fat-quality measurements but requires sophisticated statistical methods to evaluate individual data properly. The main advantage of colorimetric methods is the measurement of a single group of flavor compounds, which is important because of its sensory properties. With this in mind, a colorimetric method was developed in our laboratory to analyze volatile saturated and unsaturated carbonyls in oils and fried products. In a single run of relatively short duration, both groups of flavor compounds can be measured by simple colorimetry (65). These results can be directly transferred to oil quality and excellent correlations were found with sensory evaluation (66).

## Database

The majority of the volatile components formed from oxidized oils and fats are directly related to the fatty acid composition and the presence of minor components that stimulate oxidation. Data libraries or databases must include this information in order to assess possible volatiles formed. Data collected and/or obtained from published literature should include source of volatiles, oil/fat type, oxidation level, thermal treatment during storage and volatiles analysis, exposure to light, and any other factors that can affect the stability of hydroperoxides and the oxidation rate of lipids. To build a database related to volatiles, two types of information should be included: qualitative data describing the type of compounds by chemical structure, function, and sensory properties, including sensory characterization as dependent on concentration; and quantitative data on the concentrations of volatiles in particular oils or fats at certain levels of oxidation or deterioration.

The quantitative data combined with sensory characteristics can be very useful for quality assessment of oils and fats having a similar fatty acid composition. Both groups of data can be used to predict the type of volatiles formed from oils and fats with a particular fatty acid composition and their effect on sensory characteristics. This data can also be utilized to construct pattern(s) of volatiles formation for (particular) oils or fats to be used for quality control purposes.

## Conclusion

Based on the studies discussed, it is clear that there are differences between the methods for volatile analysis and the volatile markers used for assessing lipid oxidation of different oils. In addition, the particular method used can also exert considerable influence on the volatiles identified and quantified. In an attempt to build a suitable volatile database, it is important to establish the optimal and most consistent method for analysis of the volatiles actually present in the oil. Of the techniques developed, MHE and direct gas chromatography appear to be particularly promising, although the conditions used should be close to those for sensory assessment. Carrying out both analyses under similar conditions will produce more useful correlations. Some of the volatiles discussed could prove to be useful markers in oil quality assessment. Less abusive methods are essential to minimize the formation of artifacts and additional volatiles during analytical procedures.

### References


1. Frankel, E.N. 1980. *Prog. Lipid Res. 19:*1.
2. Forrs, D.A. 1972. *Prog. Chem. Fats Other Lipids 13:*181.
3. Schieberle, P., and Grosch, W. 1981. *J. Am. Oil Chem. Soc. 58:*602.
4. Eriksson, C.E. In *Autoxidation of Unsaturated Lipids.* Edited by H.W.-S. Chan, Academic Press, London, 1987, p. 207.
5. Holm, U., Ekbom, K., and Wade, G. 1957. *J. Am. Oil Chem. Soc. 34:*606.
6. Frankel, E.N. 1981. *Prog. Lipid Res. 22:*1.
7. Grosch, W. In *Autoxidation of Unsaturated Lipids.* Edited by H.W.-S. Chan, Academic Press, London, 1987, p. 95.
8. Josephson, D.B., and Lindsay, R.C. 1987. *J. Am. Oil Chem. Soc. 64:*132.
9. Cooney, P.M., Evans, C.D., Schwab, A.W., and Cowan, J.C. 1958. *J. Am. Oil Chem. Soc. 35:*152.
10. Min, D.B. 1983. *J. Am. Oil Chem. Soc. 60:*544.
11. Pokorny, J. In *Flavor Chemistry of Lipid Foods.* Edited by D.B. Min and T.H. Smouse, The American Oil Chemists' Society, Champaign, IL, 1989, p. 113.
12. Rothe, M. 1980. *Nahrung 24:*7.
13. Heath, H.B., and Reineccius, G. *Flavor Chemistry and Technology*, AVI Publishing, Westport, 1986, p. 71.

14. Min, D.B., and Schweizer, D.Q. 1983. *J. Am. Oil Chem. Soc. 60:*1662.
15. Waltking, A.E., and Zmachinski, H. 1977. *J. Am. Oil Chem. Soc. 54:*454.
16. Dupuy, H.P., Rayner, E.T., Wadsworth, J.I., and Legendre, M.C. 1977. *J. Am. Oil Chem. Soc. 54:*445.
17. Selke, E., and Rohwedder, W.K. 1983. *J. Am. Oil Chem. Soc. 60:*1853.
18. Snyder, J.M., and Mounts, T.L. 1990. *J. Am. Oil Chem. Soc. 67:*800.
19. Frankel, E.N. In *Flavor Chemistry of Lipid Foods.* Edited by D.B. Min and T.H. Smouse, The American Oil Chemists' Society, Champaign, IL, 1989, p. 32.
20. Frankel, E.E., Selke, E., Snyder, J.M., and Warner, K. 1988. *J. Am. Oil Chem. Soc. 65:*1617.
21. Buttery, R.G., Guadagni, D.G., and Ling, L.C. 1973. *J. Agric. Food Chem. 21:*198.
22. Dixon, M.D., and Hammond, E.G. 1984. *J. Am. Oil Chem. Soc. 61:*1452.
23. Malcolmson, L., Przybylski, R., and Ryland, D. 1993. *INFORM 4:*378.
24. Min, D.B. 1981. *J. Food Sci. 46:*1453.
25. Rayner, E.T., Wadsworth, J.I., Legendre, M.G., and Dupuy, H.P. 1978. *J. Am. Oil Chem. Soc. 55:*454.
26. Henderson, S.K., and Nawar, W.W. 1981. *J. Am. Oil Chem. Soc. 58:*632.
27. Vasundhara, T.S., and Parihar, D.B. 1980. *Lebensm. Wiss. Technol. 13:*78.
28. Sessa, D.J., and Plattner, R.D. 1979. *J. Agric. Food Chem. 27:*209.
29. Yajima, I., Nakamura, M., Sakakibara, H., Yanai, T., and Hayashi, K. 1981. *Agric. Biol. Chem. 45:*2761.
30. Waltking, A.E., and Goetz, A.G. 1983. *CRC Critr. Rev. Food Sci. Nutr. 19:*99.
31. Mordret, F., Morin, O., and Coustille, J.L. 1985. *Revue Francaise Crop Grass. 32:*193.
32. Frankel, E.N. 1993. *Trends Food Sci. Technol. 5:*456.
33. Ioffe, B.V., and Vitenberg, A.G. *Head-Space Analysis and Related Methods in Gas Chromatography.* John Wiley & Sons, New York, 1983, p. 9.
34. Venema, A. 1990. *J. High Res. Chrom. 13:*537.
35. Warner, K., and Frankel, E.N. 1985. *J. Am. Oil Chem. Soc. 62:*100.
36. Snyder, J.M., Frankel, E.N., Selke, E., and Warner, K. 1988. *J. Am. Oil Chem. Soc. 65:*1617.
37. Frankel, E.N., Nash, A.M., and Snyder, J.M. 1987. *J. Am. Oil Chem. Soc. 64:*987.
38. Snyder, J.M., Frankel, E.N., and Warner, K. 1986. *J. Am. Oil Chem. Soc. 63:*1055.
39. Matthews, R.F., Scanlan, R.A., and Libbey, L.M. 1971. *J. Am. Oil Chem. Soc. 48:*745.
40. Murray, K.E. 1977. *J. Chrom. 135:*49.
41. Kolb, B. 1982. *Chromatographia 15:*587.
42. Raghavan, S.K., Reeder, S.K., and Khayat, A. 1989. *J. Am. Oil Chem. Soc. 66:*942.
43. Simpson, R. 1979. *Chromatographia 12:*733.
44. Rapp, A., and Knipser, W. 1980. *Chromatographia 13:*698.
45. Nunez, A.J., Gonzales, L.F., and Janak, J. 1984. *J. Chrom. 300:*127.
46. Przybylski, R., Eskin, N.A.M., and Malcolmson, L. 1991. *Can. Inst. Food Sci. Technol. J. 24:*129.
47. Bianchi, A., Varney, M.S., and Philips, J. 1989. *J. Chrom. 467:*111.
48. Hinshaw, J.V. 1990. *LC-GC 6:*873.
49. MacLeod, G., and Ames, J.M. 1986. *J. Chrom. 355:*393.
50. Midleditch, B.S., and Zlatkis, A. 1987. *J. Chrom. Sci. 25:*547.
51. Legendre, M.G., Fisher, G.S., Schuller, W.H., Dupuy, H.P., and Rayner, E.T. 1979. *J. Am. Oil Chem. Soc. 56:*552.

52. Legendre, M.G., Dupuy, H.P., Rayner, E.T., and Schuller, W.H. 1980. *J. Am. Oil Chem. Soc. 57:*361.
53. Snyder, J.M., Frankel, E.N., and Selke, E. 1985. *J. Am. Oil Chem. Soc. 62:*1657.
54. Scholz, R.G., and Ptak, L.R. 1966. *J. Am. Oil Chem. Soc. 43:*596.
55. Warner, K., and Frankel, E.N. 1985. *J. Am. Oil Chem. Soc. 62:*100.
56. Przybylski, R. In *Proceedings of the Eighth International Congress.* Edited by D.I. McGregor, Saskatoon, Canada, 1991, p. 861.
57. Jarvi, P.K., Lee, G.D., Erikson, D.R., and Butkus, E.A. 1971. *J. Am. Oil Chem. Soc. 48:*121.
58. Warner, K., Evans, C.D., List, G.R., Boundy, B.K., and Kwolek, W.J. 1974. *J. Food Sci. 39:*761.
59. Morrison III, W.D., Lyon, B.G., and Robertson, J.A. 1981. *J. Am. Oil Chem. Soc. 58:*23.
60. Dupuy, H.P., Fore, S.P., and Goldblatt, L.A. 1971. *J. Am. Oil Chem. Soc. 48:*876.
61. Buttery, R.G., and Teranishi, R. 1963. *J. Agric. Food Chem. 11:*504.
62. Fritsch, C.W., and Gale, J.A. 1977. *J. Am. Oil Chem. Soc. 54:*225.
63. Warner, K., Evans, C.D., List, G.R., Dupuy, H.P., Wadsworth, J.I., and Goheen, G.E. 1978. *J. Am. Oil Chem. Soc. 55:*252.
64. Robards, K., Kerr, A.F., Patsalides, E., and Korth, J. 1988. *J. Am. Oil Chem. Soc. 65:*1621.
65. Przybylski, R., and Hougen, F. 1989. *J. Am. Oil Chem. Soc. 66:*1465.
66. Przybylski, R., Malcolmson, L., Eskin, N.A.M., Durance-Tod, S., Mickle, J., and Carr, R. 1993. *Lebensm. Wiss. Technol. 26:*205.
67. Frankel, E.N., Selke, E., Snyder, J., and Warner, K. 1988. *J. Am. Oil Chem. Soc. 65:*1617.
68. Forss, D.A. 1972. *Prog. Chem. Fats and Other Lipids 13:*181.

Chapter 7

# Historical and Future Development of Volatile Compound Analyses

**Janet M. Snyder**

Food Quality and Safety Research, 1815 N. University St., Peoria, IL 61604

## Introduction

Flavor volatiles from seed oils have been extensively researched, and many gas chromatographic methods have been developed to characterize the compounds formed (1). Lipid oxidation can produce many heat-labile compounds, and the sample matrix can change during analysis unless mild conditions are used (2-4). Quantitation methods for determining the concentration of volatiles vary, making reproducibility difficult. Methods that are developed for other matrices, such as solid or aqueous samples, are not always effective for analyzing oil volatiles. This chapter reviews the development of techniques for volatile analysis, including steam and vacuum distillation, dynamic headspace, direct injection, and static headspace, from the early work in the 1960s to the present. Descriptions and applications of each method are presented, and advantages and disadvantages are reported. Future technology, such as supercritical methods, is also discussed.

Gas chromatographic (GC) analysis for oil volatiles was developed in the early 1960s; Nawar (5) injected volatile compounds into a GC using a closed circulating system to concentrate the volatile compounds from corn oil and other foods. The development of several methods for qualitative and quantitative analysis of oil volatiles included distillation, dynamic headspace, static headspace, and direct injection.

## Distillation

Concentration of the volatile compounds required for analysis of the trace components formed from lipids was initially accomplished by distillation. Chang reported a new technique to isolate and concentrate flavor compounds by steam distillation in 1961 (6). Soybean oil and corn oil were aged; volatiles were collected by distillation and then analyzed by GC. Vacuum distillation as described by de Bruyn and Schogt (7) has been used and modified by many researchers. Decomposition products from oxidized soybean oil were collected by vacuum distillation and identified by gas chromatography—mass spectrometry (GC–MS) by Selke and co-workers (8). They were the first to use a GC–MS system to separate and identify oxidation products from soybean oil. Adaptations of the vacuum-distillation technique have been used to identify trace components in hydrogenated peanut oil (9) and oxidation products from fatty acid components of vegetable oils (10).

## Dynamic Headspace

Vacuum distillation techniques were modified by collecting volatiles by cryogenic trapping or using a cold finger (11–16). Nawar and co-workers reported the development of a distillation system for efficient recovery of lipid volatiles (11). Selke et al. designed an apparatus to collect and identify volatiles from heated oils and described the reactor as a "microroom" (Figure 7.1 [12]). Air was swept over the heated oil directly into a GC column, cryogenically cooled to –60°C for dynamic

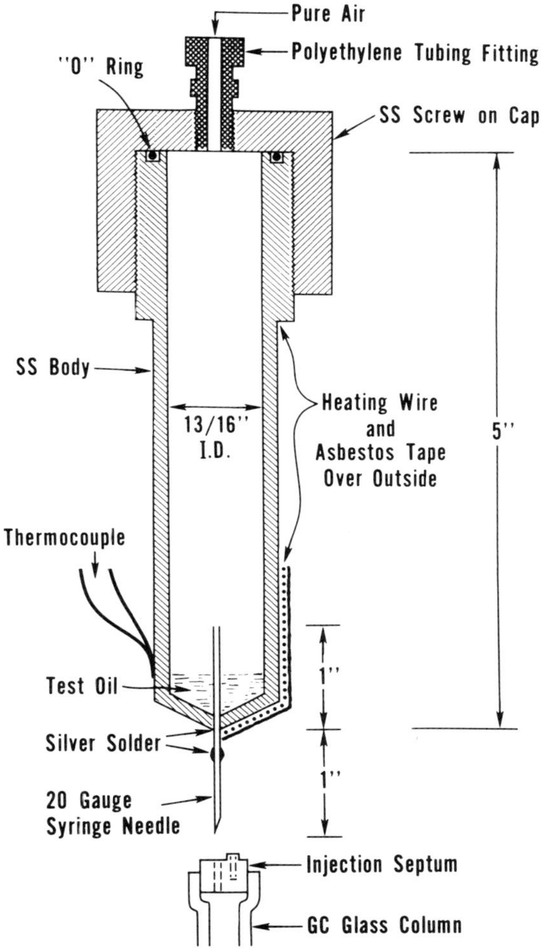


**Figure 7.1.** Stainless steel reactor (microroom) designed to simulate deep fat frying temperature and to collect odors from 1 to 5 mL of heated soybean oil directly into a gas chromatographic column (12).

concentration of the volatile components, and the volatiles were analyzed by GC-MS. This procedure was used to identify many compounds formed from tristearin (13), triolein (14), trilinolein (15), and trilinolenin (16). The purge and trap method was evolved in order to preconcentrate trace compounds. Oil samples were analyzed by sweeping the oil with nitrogen or an inert gas and collecting the volatiles on adsorbent material, such as Tenax (17). Similar methods were used to concentrate the volatile components and quantitatively isolate flavor components in palm oil (18).

Commercial instruments that utilized the principle of purging the oil and collecting the volatiles in porous traps soon became available. These units have been used in a variety of applications for seed oil analysis (19–23). In 1983, Roberts analyzed vegetable oil volatiles using a commercial unit (19). Volatiles, automatically sampled by sweeping the oil with an inert gas, were trapped on Tenax, a porous polymer. Even low concentrations of 1 ppb of a volatile compound from corn oil were collected and thermally desorbed into a GC column for analysis (20). A trap containing Tenax, Ambersorb (a polymeric adsorbent), and graphite was used in a dynamic headspace system to study the effect of tertiary butylhydroquinone (TBHQ) on the production of volatiles in vegetable oils (21). Selke and Frankel showed that the temperature of the oil during the purge step of an automatic system affected the concentration of individual volatiles (Figure 7.2 [4]). Automated dynamic systems have been improved to shorten analysis times and to develop multiple concentration techniques for rapid evaluation of vegetable oils (22). Engeseth et al. analyzed crude soybean volatiles by purge and trap to assess tempering of genetically modified soybeans (23).

## Direct Injection

Direct injection, the simplest method for volatile analysis requires no previous sample enrichment. In 1966, Scholz and Ptak developed a procedure for determining the amount of pentane in oil samples and comparing the results to flavor evaluation panel tests (24). Evans reported on the use of a direct injection tube for the analysis of hydrocarbons from oxidized soybean oil and linseed oil at an American Oil Chemists' Society meeting in 1967 (25). Early chromatograms from the research by Evans (Figure 7.3 [25]) showed the separation of hydrocarbons formed during autoxidation; however, the quality of the chromatograms was poor compared to today's capillary column technology. Later studies reported the correlation of pentane with thermal decomposition of oils and the relationship of pentane to flavor scores of oxidized soybean and cottonseed oils (26–27). Direct injection of the headspace volatiles was used by Warner to measure pentane for correlation with potato chip rancidity determined by sensory analysis (28).

Dupuy developed a device for direct injection in which vegetable oil was introduced into an injector containing glass wool (29). Volatile compounds were backflushed into the GC column using the carrier gas. After GC analysis, nonvolatiles

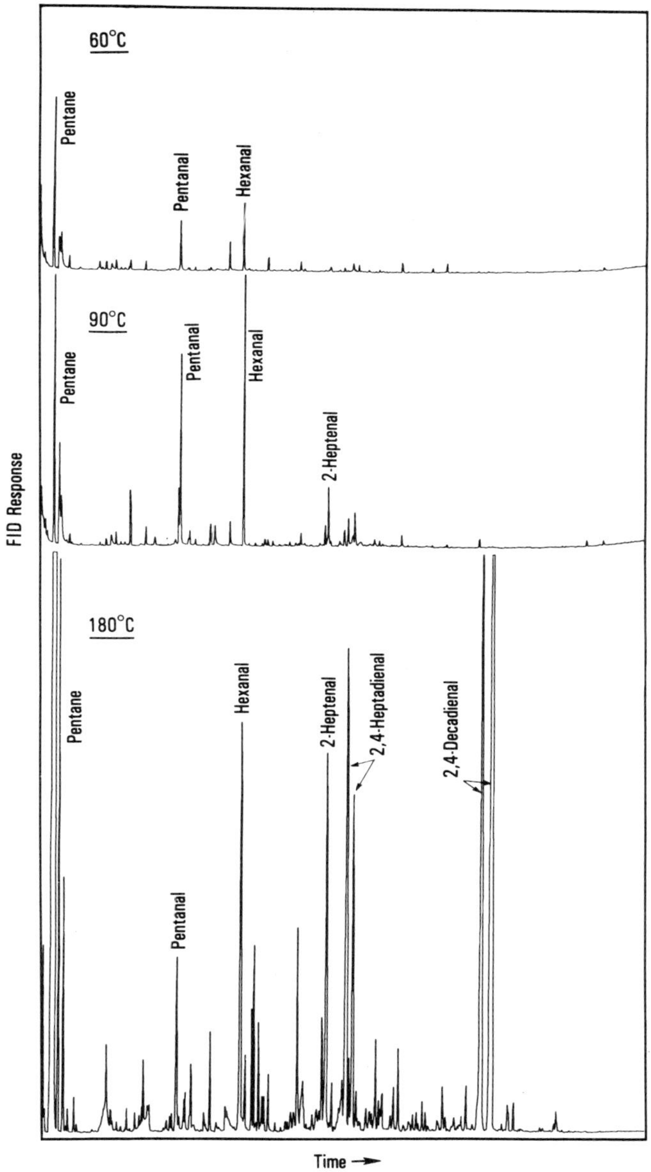


Fig. 2

**Figure 7.2.** Effect of sampling temperatures on dynamic headspace gas chromatograms of soybean oil oxidized at ambient temperature to a peroxide value of 5.3 in the presence of citric acid (4).

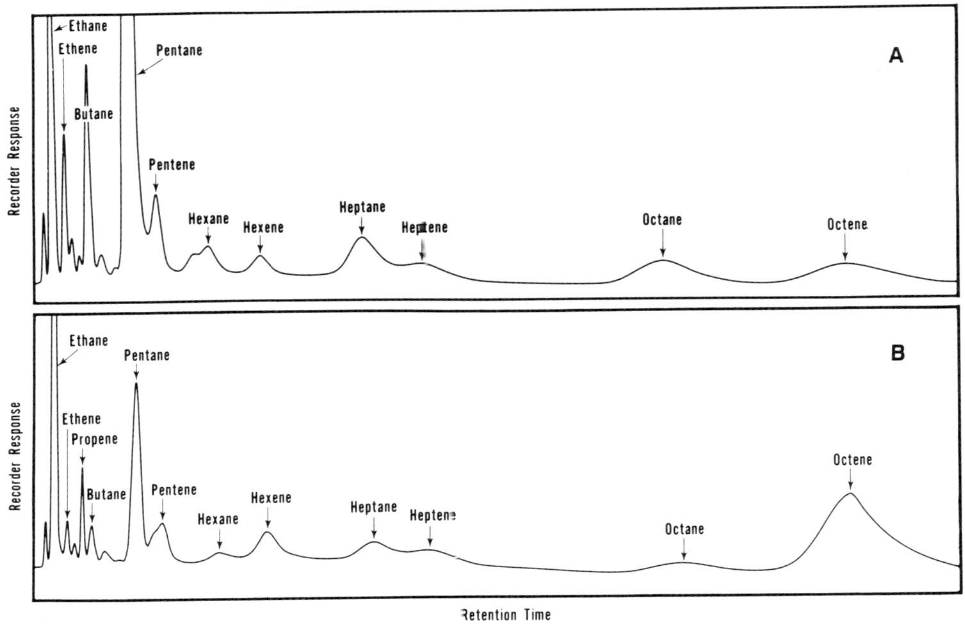


**Figure 7.3.** Early GC chromatograms of hydrocarbons from autooxidized vegetable oils. Analyses performed in 1967 (25). (a) Soybean oil. (b) Linseed oil.

were removed from the system before the insertion of another sample. This technique and modified versions were used to investigate volatiles and to identify compounds from the oxidation of edible oils (30–35). Jackson modified the direct injection procedure of Dupuy by sampling directly onto a Porapak column (36), and connecting the column to a gas chromatograph to identify and quantitate volatiles from aged soybean oil (37). Williams and Wille substituted a polymetaphenoxylene (Poly MPE)-Tenax column for the Porapak P column and separation of volatiles was improved (38). Additional modifications improved the correlation of volatile analysis with sensory data (39). Warner used a direct injection method for monitoring volatiles to predict the stability of soybean oil (40). In addition, the effect of β-carotene on the stability of photooxidized soybean oil was evaluated by the direct injection analysis of volatiles generated from oxidation products during light-exposure tests (41).

## Static Headspace

Direct injection methods were preferred by researchers because there was no preconcentration of the volatiles. Equilibrium headspace sampling or static headspace also eliminated the trapping step and became another rapid analysis method. After commercial static headspace instrumentation was marketed in 1976, methods were developed for edible oil analyses. In 1984, Marsili reported the benefits of static

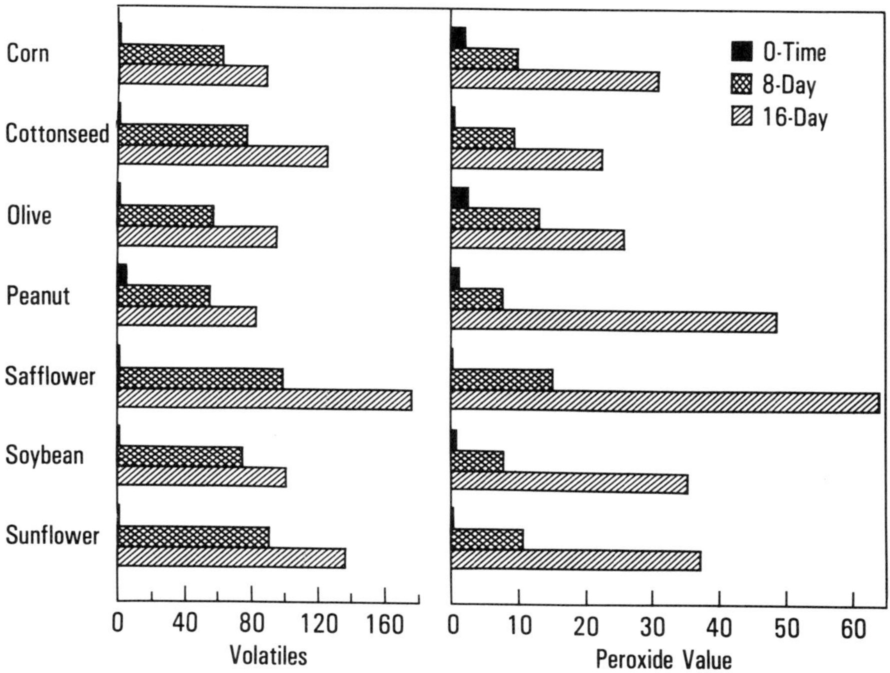


**Figure 7.4.** Comparison of peroxide value and total GC volatiles of 7 vegetable oils after storage at 60°C (42).

headspace sampling; volatile analyses of photooxidized soybean oil were reproducible by this method (42). Snyder et al. investigated static headspace sampling to measure the formation of volatile compounds after storage of eight vegetable oils (43). The relationship of total volatiles and peroxide value (PV) from the oxidized vegetable oils were reported (Figure 7.4).

The stability of untreated and hydrogenated soybean oils were further analyzed by static headspace technique to evaluate the effects of TBHQ and methyl silicone by measuring the total volatiles formed during storage (44). Quantitation of the volatile components from five edible oils was improved by a multiple extraction technique (45). The headspace from a vial containing the oil was sampled, analyzed by GC, and sampled twice more to provide a reproducible determination of individual compounds (Figure 7.5). Multiple headspace extraction was instrumental in evaluating the effects of microwave treatment on soybeans (46). Enzymes were inactivated by microwave heating, and hexanal development from the soybeans and soybean oil was monitored during an 8-week period. Static headspace analysis provided a rapid evaluation of soybean, sunflower, and canola oil stability as reported by Warner and co-workers (47). Improvements have helped establish static headspace as a reproducible method for the routine analysis of oxidized oil samples (48).

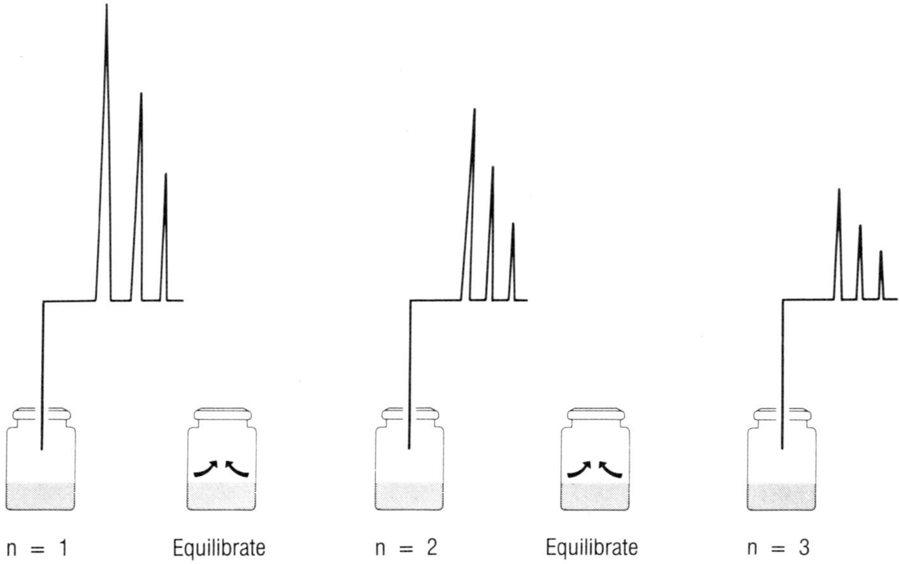


**Figure 7.5.** Multiple headspace extraction diagram. Soybean oil in sealed vial heated at 90°C for 30 min., headspace volatiles were injected ($n$ = 1). Vial was vented and equilibrated before next injection ($n$ = 2). Procedure was repeated ($n$ = 3).

## Comparison of GC Volatile Analysis Methods

Direct injection, and dynamic and static headspace methods were compared to demonstrate the differences in the volatile profiles generated by each method (49). Chromatograms presented in Figure 7.6 illustrate how each method preferentially favors certain volatile compounds. Higher molecular weight compounds, such as 2,4-decadienal isomers, were largest with direct injection. Low molecular weight compounds were most pronounced in static headspace, because of the partition equilibrium between volatiles in the oil sample and the headspace volatiles. Dynamic headspace provided good recoveries of most components; however, some lower boiling point compounds were not completely retained in the trap during the purge cycle, and the smaller pentane peak could be the result of the inefficient trapping.

Methods for volatile analysis were evaluated by Prevot in 1971 (50). More recently, Mordret et al. listed advantages and disadvantages of four volatile analysis methods, direct injection, static headspace, molecular distillation, and dynamic headspace (51). Direct injection rated highest and static headspace rated second for ease of operation, rapid analysis time, ability for routine analysis, and cost. Dynamic headspace provided the highest sensitivity followed by molecular distillation, static headspace, and direct injection. Selectivity was diminished with static headspace sampling due to the loss of heavy compounds (51). Concentrations of

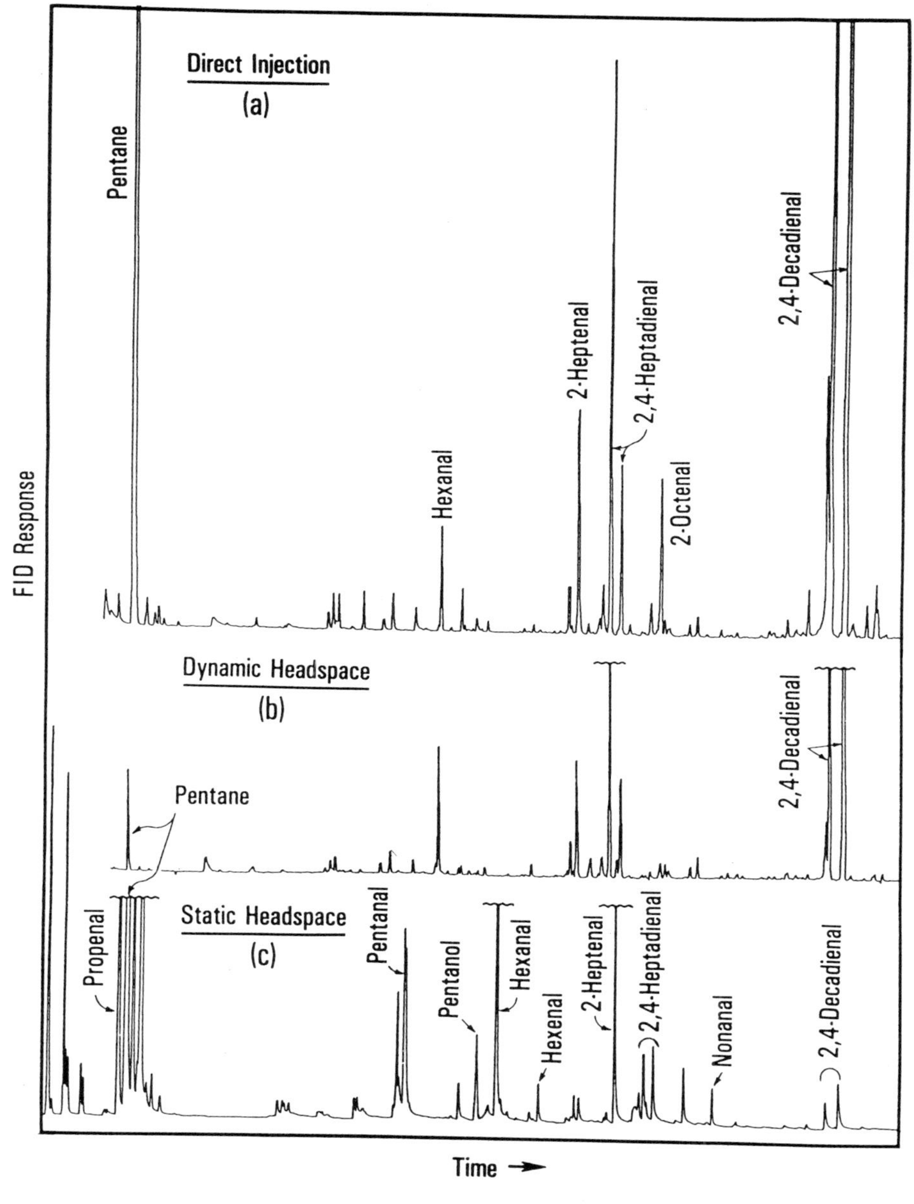


**Figure 7.6.** Comparison of three GC methods for volatile analyses of oxidized soybean oil (peroxide value = 9.5) heated to 180°C (49). (a) Direct injection. (b) Dynamic headspace. (c) Static headspace.

volatiles in the vapor phase were dependent on the partition coefficients, and concentration of low molecular weight compounds were greater. Lower selectivity was reported with molecular distillation because the lighter compounds were not trapped efficiently and lost. The memory effect of ghost peaks from sample to sample was reported to be greatest using dynamic headspace analysis; volatiles were not completely removed from tubing and dynamic systems between samples, requiring extensive clean-out procedures (51).

## The Future of Volatile Analysis

New technologies are being developed and made commercially available for more efficient handling of volatiles. The systems are more automated, and applications for seed oils are being explored to more efficiently analyze the compounds. Recently, a thermal desorption instrument was marketed for easy attachment to GC injection systems. The thermal desorption method can be used for qualitative and quantitative studies of the low concentration (ppt) volatiles generated from food and food products (52). Further work with oil volatiles and semivolatiles, and high molecular weight compounds, such as hexadecanol and fatty acids, is progressing.

Supercritical fluid methods have been developed to selectively extract flavor components from a variety of plants, spices, and food products (53–55). Concentration of volatiles can be accomplished using supercritical $CO_2$ (SC-$CO_2$) at temperatures as low as 32°C. This technology concentrates the volatile and semivolatile compounds from edible oils without degrading thermally labile compounds.

The supercritical fluid method developed for edible oils is similar to a purge and trap operation. In a study reported at the AOCS Meeting in 1992, volatile compounds from vegetable oils extracted with SC-$CO_2$ were investigated to determine if the supercritical extracted oil was comparable to hexane-extracted oil (56). During the extraction process, volatile components were concentrated in Tenax traps located at the exit port of the oil receiver. The decompressed $CO_2$ flow rate was maintained at 3 L/min. for discriminate capture of the volatiles.

Volatiles were analyzed by thermal desorption from the Tenax traps into a cryogenically cooled GC column in the same manner as a dynamic headspace sampler operates. Identification and quantitation of the volatiles was accomplished by GC–MS. The use of SC-$CO_2$ to desorb the volatiles from the Tenax traps into the GC column was also investigated. Temperature, pressure, and $CO_2$ flow rate were varied to develop the best conditions for analytical determination. The supercritical desorption method was reported as a useful technique for detecting higher molecular weight compounds and semivolatiles. Comparison of the volatile profiles from corn oil indicated the differences between thermal desorption and supercritical desorption (Figure 7.7). A higher amount of low molecular weight compounds were analyzed by thermal desorption, while the high molecular weight compounds were obtained by supercritical desorption (56).

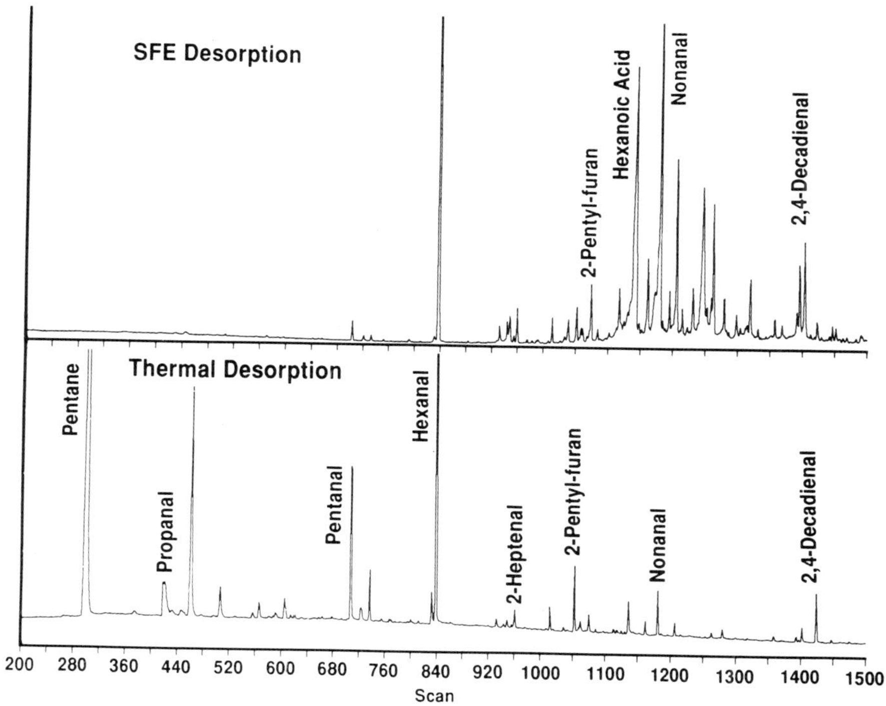


**Figure 7.7.** Comparison of supercritical and thermal desorption methods. GC-MS ion current chromatograms of supercritical fluid extracted corn oil at 8000 psi and 50°C.

Options using supercritical technologies should be explored in the future. Extraction of the volatiles directly from the oil by SC-$CO_2$ into the GC column is a promising advancement. The method is nondestructive and can be used to concentrate trace components. The goal of this research is to develop a simple, rapid method that is both qualitative and quantitative.

## References


1. Waltking, A.E., and Goetz, A.G. 1983. *CRC Crit. Rev. in Food Sci. and Nutr. 19:*99.
2. Frankel, E.N., Neff, W.E., and Selke, E. 1981. *Lipids 16:*279.
3. Nawar, W.W. *Thermal Decomposition of Lipids,* ACS Symposium Series 409, American Chemical Society, Washington, D.C., 1989, p. 114.
4. Selke, E., and Frankel, E.N. 1987. *J. Am. Oil Chem. Soc. 64:*749.
5. Nawar, W.W., and Fagerson, I.S. 1960. *Anal. Chem. 32:*1534.
6. Chang, S.S. 1961. *J. Am. Oil Chem. Soc. 38:*669.
7. de Bruyn, J., and Schogt, J.C.M. 1961. *J. Am. Oil Chem. S 38:*40.
8. Selke, E., Moser, H.A., and Rohwedder, W.K. 1970. *J. Am. Oil Chem. Soc. 47:*393.
9. Feenstra, W.H., and Meijboom, P.W. 1971. *J. Am. Oil Chem. Soc. 48:*684.

10. Henderson, S.K., Witchwoot, A., and Nawar, W.W. 1980. *J. Am. Oil Chem. Soc. 57:*409.
11. Nawar, W.W., Champagne, J.R., Dubravic, M.F., and LeTellier, P.R. 1969. *J. Agric. Food Chem. 17:*645.
12. Selke, E., Rohwedder, W.K., and Dutton, H.J. 1972. *J. Am. Oil Chem. Soc. 49:*636.
13. Selke, E., Rohwedder, W.K., and Dutton, H.J. 1975. *J. Am. Oil Chem. Soc. 52:*232.
14. Selke, E., Rohwedder, W.K., and Dutton, H.J. 1977. *J. Am. Oil Chem. Soc. 54:*62.
15. Selke, E., Rohwedder, W.K., and Dutton, H.J. 1980. *J. Am. Oil Chem. Soc. 57:*25.
16. Selke, E., and Rohwedder, W.K. 1983. *J. Am. Oil Chem. Soc. 60:*1853.
17. Singleton, J.A., and Pattee, H.E. 1980. *J. Am. Oil Chem. Soc. 57:*405.
18. Dirinck, P.J., Kuntom, A.H., and Schamp, N. In *Progress in Flavour Research 1984,* edited by J. Adda, Elsevier Science Publishers B.V. Amsterdam, 1985, p. 505.
19. Roberts, J. "Semiautomated Dynamic Headspace Analysis of Vegetable Oil Volatiles" presented at the 74th Meeting of the American Oil Chemists' Society, Chicago, IL, May, 1983.
20. Westendorf, R.G. In *Characterization and Measurement of Flavor Compounds,* edited by D.D. Bills and C.J. Mussinan, ACS Symposium series 289, American Chemical Society, Washington, D.C., 1985, p. 138.
21. Wyatt, D.M. 1987. *J. Chromatog. Sci. 25:*257.
22. Raghavan, S.K., Reeder, S.K., and Khayat, A. 1989. *J. Am. Oil Chem. Soc. 66:*942.
23. Engeseth, N.J., Perkins, E.G., and Klein, B.P. 1992. *J. Am. Oil Chem. Soc. 69:*841.
24. Scholz, R.G., and Ptak, L.R. 1966. *J. Am. Oil Chem. Soc. 43:*596.
25. Evans, C.D. "Hydrocarbons Derived from Autoxidized Vegetable Oils Then Thermal Splitting," presented at the American Oil Chemists' Society Meeting, Chicago, IL, October 15–18, 1967.
26. Evans, C.D., List, G.R., Dolev, A., McDonnell, D.G., and Hoffman, R.L. 1967. *Lipids 2:*432.
27. Evans, C.D., List, G.R., Hoffman, R.L., and Moser, H.A. 1969. *J. Am. Oil Chem. Soc. 46:*501.
28. Warner, K., Evans, C.D., List, G.R., Boundy, K., and Kwolek, W.F. 1974. *J. Food Sci. 39:*761.
29. Dupuy, H.P., Fore, S.P., and Goldblatt, L.A. 1971. *J. Am. Oil Chem. Soc. 48:*876.
30. Dupuy, H.P., Fore, S.P., and Goldblatt, L.A. 1973. *J. Am. Oil Chem. Soc. 50:*340.
31. Waltking, A.E., and Zmachinski, H. 1977. *J. Am. Oil Chem. Soc. 54:*454.
32. Legendre, M.G., Fisher, G.S., Schuller, W.H., Dupuy, H.P., and Rayner, E.T. 1979. *J. Am. Oil Chem. Soc.* 56:552.
33. Warner, K., Evans, C.D., List, G.R., Dupuy, H.P., Wadsworth, J.I., and Goheen, G.E. 1978. *J. Am. Oil Chem. Soc. 55:*252.
34. Dupuy, H.P., Rayner, E.T., Wadsworth, J.I., and Legendre, M.G. 1977. *J. Am. Oil Chem. Soc. 54:*445.
35. Dupuy, H.P., Flick, G.J., Jr., Bailey, M.E., St. Angelo, A.J., Legendre, M.G., and Sumrell, G. 1985. *J. Am. Oil Chem. Soc. 62:*1690.
36. Jackson, H.W., and Giacjero, D.J. 1977. *J. Am. Oil Chem. Soc. 54:*458.
37. Jackson, H.W. 1981. *J. Am. Oil Chem. Soc. 58:*227.
38. Williams, J.L., and Wille, J.H. 1976. *J. Am. Oil Chem. Soc. 53:*634.
39. Min, D.B. 1981. *J. Food Sci. 46:*1453.
40. Warner, K., and Frankel, E.N. 1985. *J. Am. Oil Chem. Soc. 62:*100.

41. Warner, K., and Frankel, E.N. 1987. *J. Am. Oil Chem. Soc. 64:*213.
42. Marsili, R.T. 1984. *J. Chromatog. Sci. 22:*61.
43. Snyder, J.M., Frankel, E.N., and Selke, E. 1985. *J. Am. Oil Chem. Soc. 62:*1675.
44. Snyder, J.M., Frankel, E.N., and Warner, K. 1986. *J. Am. Oil Chem. Soc. 63:*1055.
45. Snyder, J.M., and Mounts, T.L. 1990. *J. Am. Oil Chem. Soc. 67:*800.
46. Snyder, J.M., Mounts, T.L., and Holloway, R.K. 1990. *J. Am. Oil Chem. Soc. 68:*744.
47. Warner, K., Frankel, E.N., and Mounts, T.L. 1989. *J. Am. Oil Chem. Soc. 66:*558.
48. Ulberth, F., and Roubicek, D. 1993. *Food Chem. 46:*137.
49. Snyder, J.M., Frankel, E.N., Selke, E., and Warner, K. 1988. *J. Am. Oil Chem. Soc. 65:*1617.
50. Prevot, A. 1971. *Rev. Fr. Corps Gras 18:*517.
51. Mordret, F., Morin, O., and Coustille, J.-L. 1985. *Rev. Fr. Corps Gras 32:*193.
52. Hartman, T.G., Overton, S.V., Manura, J.J., Baker, C.W., and Manos, J.N. 1991. *Food Technol. 45:*104.
53. Krukonis, V.J. In *Characterization and Measurement of Flavor Compounds,* edited by D.D. Bills and C.J. Mussinan, ACS Symposium Series 289, American Chemical Society, Washington, D.C., 1985, p. 154.
54. Moyler, D.A. In *Distilled Beverage Flavour,* edited by J.R. Piggott and A. Paterson, Ellis Horwood Ltd., Chichester, England, 1989, p. 319.
55. Hawthorne, S.B., Krieger, M.S., and Miller, D.J. 1988. *Anal. Chem. 60:*472.
56. Snyder, J.M., and King, J.W. 1992. *INFORM 3:*500.

Chapter 8

# Analyses of Peroxide Values and Headspace Oxygen

**Tae Shik Hahm and David B. Min**

Department of Food Science and Technology, The Ohio State University, Columbus, OH 43210, USA.

## Introduction

Oxidation can result in lipids becoming rancid, and is a major cause of food deterioration and loss of consumer acceptability. Microbial and enzymatic deterioration can be controlled by storing foods at a low temperature; however, this storage does not prevent lipid oxidation due to low activation energy. Therefore it is very important to understand the principles and procedures of analytical methods used to evaluate flavor quality and oxidative stability of fats and oils. This chapter will describe peroxide value determination and headspace oxygen analysis to evaluate the oxidative stability of fats and oils.

## Peroxide Value Analysis

### *Hydroperoxide Formation*

Hydroperoxide formation is a preliminary step in the oxidative rancidity of fats and oils. Therefore, the peroxide value (PV) within certain limits, can be used as an index of the oxidative state of an oil. It must be kept in mind that peroxides generally are intermediate products of lipid oxidation and lead to the formation of the volatile compounds responsible for the rancid flavor of oils. Several excellent papers on the chemical mechanisms of lipid oxidation have been published (1–6). Hydroperoxides (ROOH) are the major initial reaction products of lipid oxidation and are formed by the reaction between oils and triplet or singlet oxygen.

Triplet oxygen is a diradical molecule and only reacts with radical compounds by a free radical mechanism as shown in Figure 8.1. The hydrogen on a carbon atom adjacent to a double bond is most labile due to the nature of the electron distribution in the molecule. A free radical is formed by the removal of a labile hydrogen. The hydroperoxides formed by triplet oxygen with linoleic acid are always conjugated hydroperoxides.

Singlet oxygen, which is a nonradical and electrophilic molecule, can react directly with nonradical, electron-rich, double bond containing compounds, such as oleic, linoleic, and linolenic acids, by a nonradical mechanism as shown in Figure 8.2. The hydroperoxides formed by linoleic acid and singlet oxygen are a mixture of conjugated and nonconjugated hydroperoxides (Figure 8.2).

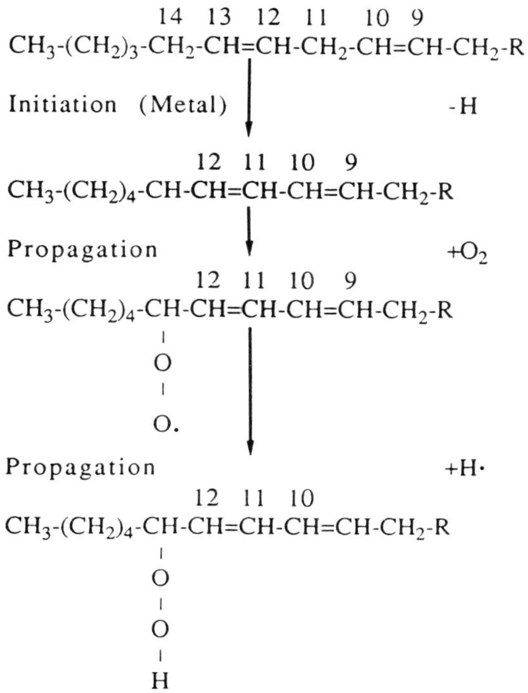


**Figure 8.1.** Formation of hydroperoxides of linoleic acid with triplet oxygen.

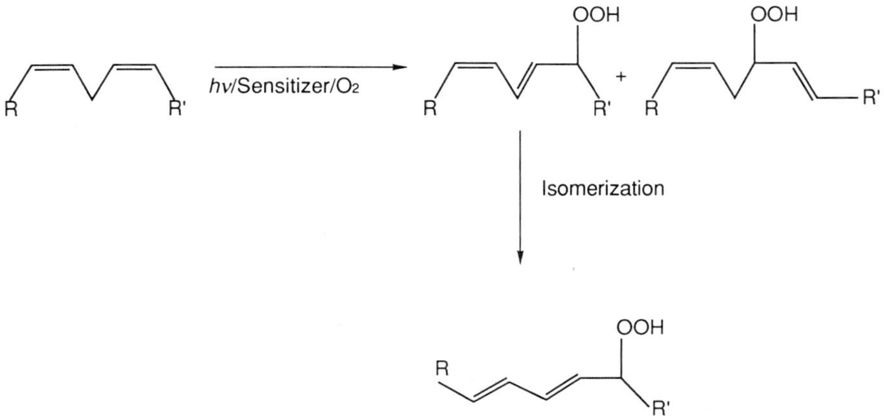


**Figure 8.2.** Formation of hydroperoxides of linoleic acid with singlet oxygen.

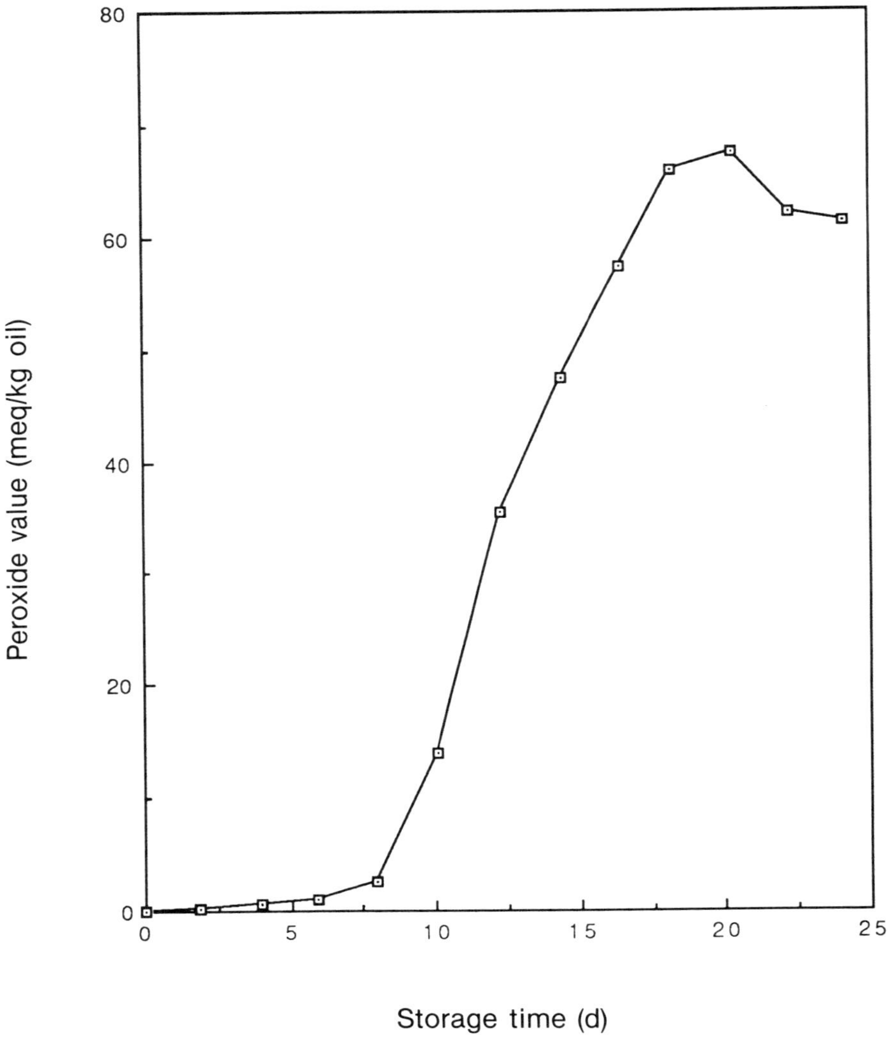


**Figure 8.3.** Slope and rate of hydroperoxide formation of stored soybean oil.

The hydroperoxides formed in the initial lipid oxidation are generally referred to as peroxides in PV determinations (7). The slope and rate of the hydroperoxide curve increases as oxidation progresses as shown in Figure 8.3 (8). Hydroperoxide formation is an autocatalytic reaction and has rate increases as a fraction of time due to the formation of products that catalyze the reaction.

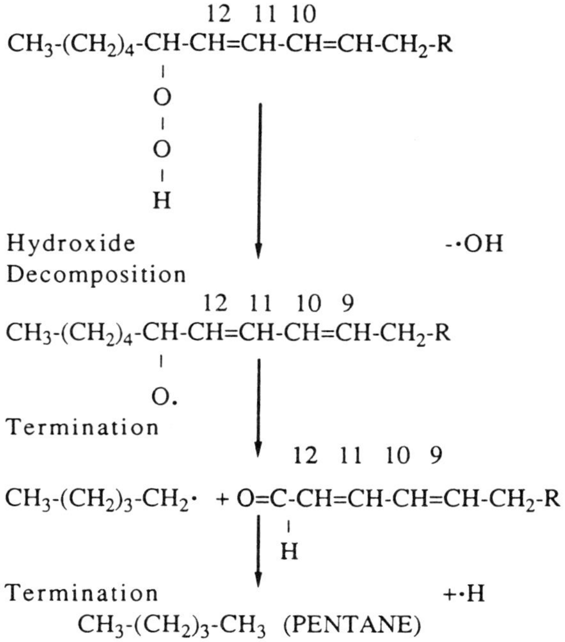


**Figure 8.4.** Hydroperoxide decomposition to form pentane and aldehydes.

### *Decomposition of Hydroperoxides*

Hydroperoxides are decomposed to low molecular weight volatile compounds, such as aldehydes, ketones, alcohols, acids, and hydrocarbons. These volatile compounds are responsible for the rancid flavor in the oxidized oils. One hundred fifty volatile compounds have been identified in oxidized vegetable oils. The mechanisms of hydroperoxide decomposition to form pentane and aldehydes are shown in Figure 8.4 (9).

### *Relationships Among Oxidized Compounds Formed During Storage*

Hydroperoxides are colorless, odorless, tasteless, heat labile, intermediate products in the formation of volatile carbonyl, hydroxyl, and hydrocarbon compounds in lipid oxidation (9). The relationships between the oxygen-containing functional groups formed during storage are shown in Figure 8.5 (10). At 55°C, hydroperoxide oxygen rose fairly steadily to a maximum of 220 milliequivalent (meg) per 100 g of oil at about 800 h before decreasing. At 100°C, the maximum peroxide oxygen was only 176 meg/100 g in about 12 h before decreasing rapidly. The hydroperoxide oxygen value leveled off as the storage period increased, indicating the rate of hydroperoxide formation is similar to the rate of hydroperoxide decomposition.

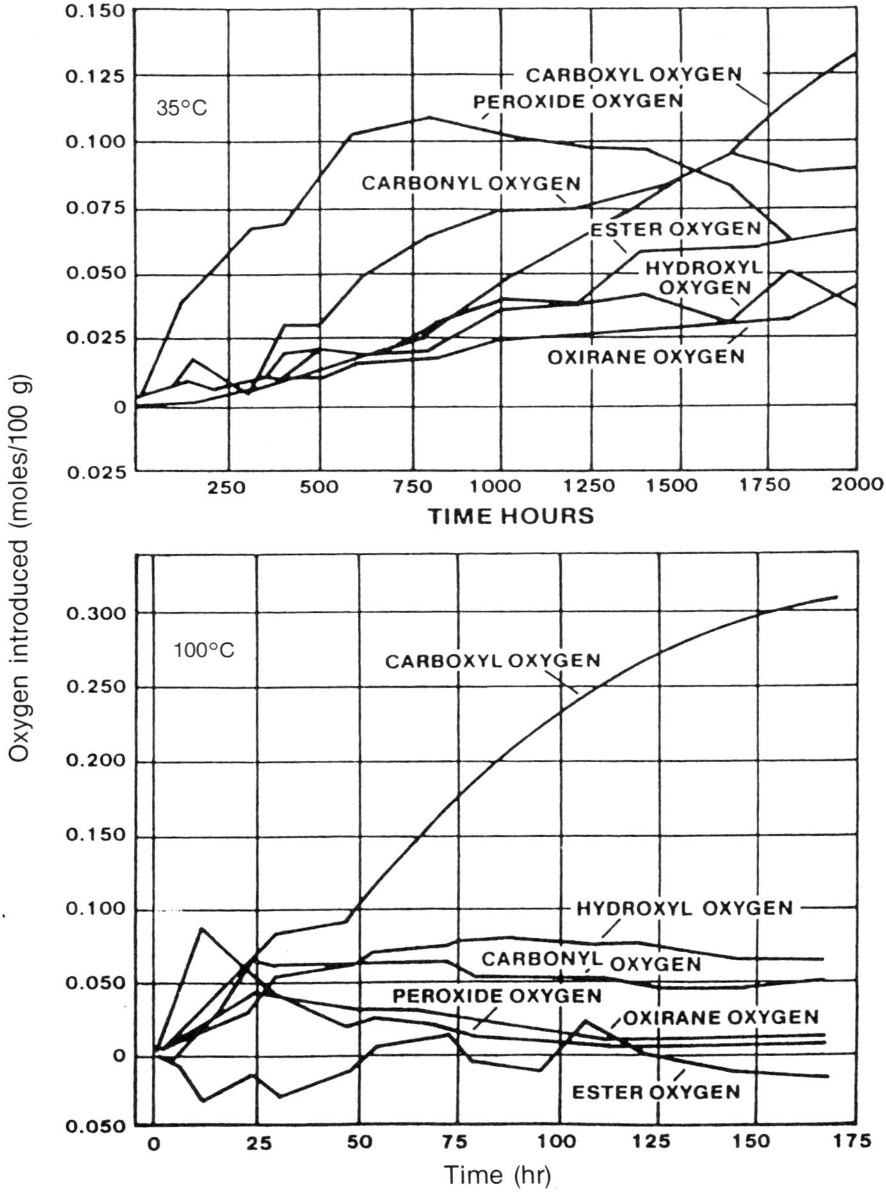


**Figure 8.5.** Relationships among oxidized compounds formed during storage.

## *Determination of PV*

Several analytical procedures for PV determination are described in literature. The results and suitability of the test depend on the experimental conditions and the reducing agent used. Whenever PV is used to evaluate the oxidative state of an oil, a detailed description of the analytical procedures used in the determination should be given. Peroxide determination methods are iodometric titration method (11–13), polarographic method, and spectrophotometric method. The widely used iodometric method is based on the measurement of the iodine produced from potassium iodide by the hydroperoxides present in the oil. Peroxide values determined by iodometric titration methods measure only the hydroperoxides in oil.

*Iodometric Peroxide Determination*

$$2KI + 2CH_3COOH \longrightarrow 2HI + 2CH_3COO\text{–}K^+$$

$$ROOH + 2HI \longrightarrow ROH + H_20 + I_2$$

$$I_2 + 2Na_2S_2O_3 \longrightarrow 2NaS_20_3 + 2NaI$$

where:

PV (meq)/1000 g sample = $(S - B)(N)$ 1000/g of sample

$S$ = Titration of sample

$B$ = Titration of blank

$N$ = Normality of sodium thiosulfate solution

The peroxide value of a fat is expressed as milliequivalents of oxygen of hydroperoxides per 1000 grams of fat (14).

## *Factors Affecting PV*

Two principal sources of error in iodometric titration method are the absorption of iodine at unsaturated bonds of fats and the liberation of iodine from potassium iodide by oxygen present in the sample solution (15). The latter is often referred to as the oxygen error and leads to high PV levels (16). Lea (11) tried to eliminate this error by filling the sample with nitrogen at the beginning of the test. Wheeler (12) used a homogeneous solution in an attempt to eliminate the need for shaking, thereby minimizing the introduction of oxygen into the sample. Other possible sources of error in the iodometric methods include variation in sample size, the type and grade of solvent used, and variation in the composition and reactivity of hydroperoxides of the sample (16). For the iodometric peroxide determination method, 91% of the oxygen absorbed by methyl linoleate was shown to be converted to peroxy compounds. The liberated iodine reacted with the unsaturated fatty acids or the oxidation products to account for the low value. The entire reaction system should be equilibrated to 37°C, and the sample should be free from pro- and antioxidants and oxygen should be rigidly excluded during the peroxide measurement.

The American Oil Chemists' Society iodometric method Cd 8-53 (13) for PV determination is applicable to all fats and oils. This commonly used method is high-

**TABLE 8.1**
Coefficients of Variation for the Analyses of Headspace Oxygen Content and Peroxide Value of Replicates of a Soybean Oil Sample

| Sample | Headspace Oxygen Content (%) | Peroxide Value (meq/kg) |
|---|---|---|
| 1 | 20.635 | 1.36 |
| 2 | 20.667 | 1.39 |
| 3 | 20.718 | 1.36 |
| 4 | 20.624 | 1.36 |
| 5 | 20.727 | 1.35 |
| 6 | 20.734 | 1.37 |
| 7 | 20.722 | 1.36 |
| 8 | 20.528 | 1.35 |
| 9 | 20.699 | 1.33 |
| 10 | 20.513 | 1.36 |
| Mean | 20.657 | 1.36 |
| SD | 0.081 | 0.015 |
| Coefficient of Variation (%) | 0.39 | 1.10 |

ly empirical and any variation in procedure may affect results. Fiedler modified the method to increase its sensitivity at low peroxide values by replacing the titration step with an electrochemical technique in which the liberated iodine was reduced on a platinum electrode maintained at a constant potential (17). Peroxide values ranging from 0.06 to 20 have been determined using this method (16). It is essential to deaerate all the solutions, since the presence of oxygen can lead to further formation of peroxides. The coefficient of variation for the determination of peroxide values of oil by the AOCS method is less than 1% as shown in Table 8.1 (8).

### *Relationship Between PV and Other Methods*

Common methods to determine the oxidative state of oils are PV, volatile compounds, measurement of absorbance at 233 nm to determine conjugated double bonds, and sensory analysis. The relationship between PV and absorbance at 233 nm, volatile compounds, and sensory scores of oils during initial oil oxidation are shown in Figures 8.6 and 8.7, and Table 8.2, respectively. Correlation coefficients between PV and other methods were 0.80 or higher. This indicates that PV can be used as an analytical method to evaluate the oxidative stability of an oil independently or to complement other methods during the initial oxidation state of an oil. Correlation coefficients of the relationship between PV and headspace oxygen content of an oil during initial oxidation of the oil was 0.87 (Figure 8.8). As the headspace oxygen of oil decreased, the peroxide value increased as expected.

The peroxide value is the most common method used to evaluate the oxidative quality of oils and fats. However, the use of PV is limited to the initial stages of oxidation. Since peroxides are vulnerable to further reaction (14), the complete oxidative history of the oil may not be revealed. Good correlations have been reported between PV and flavor scores for various commercial fats such as lard, hydrogenated soybean oil, and corn oil during the initial stages of oxidation (21).

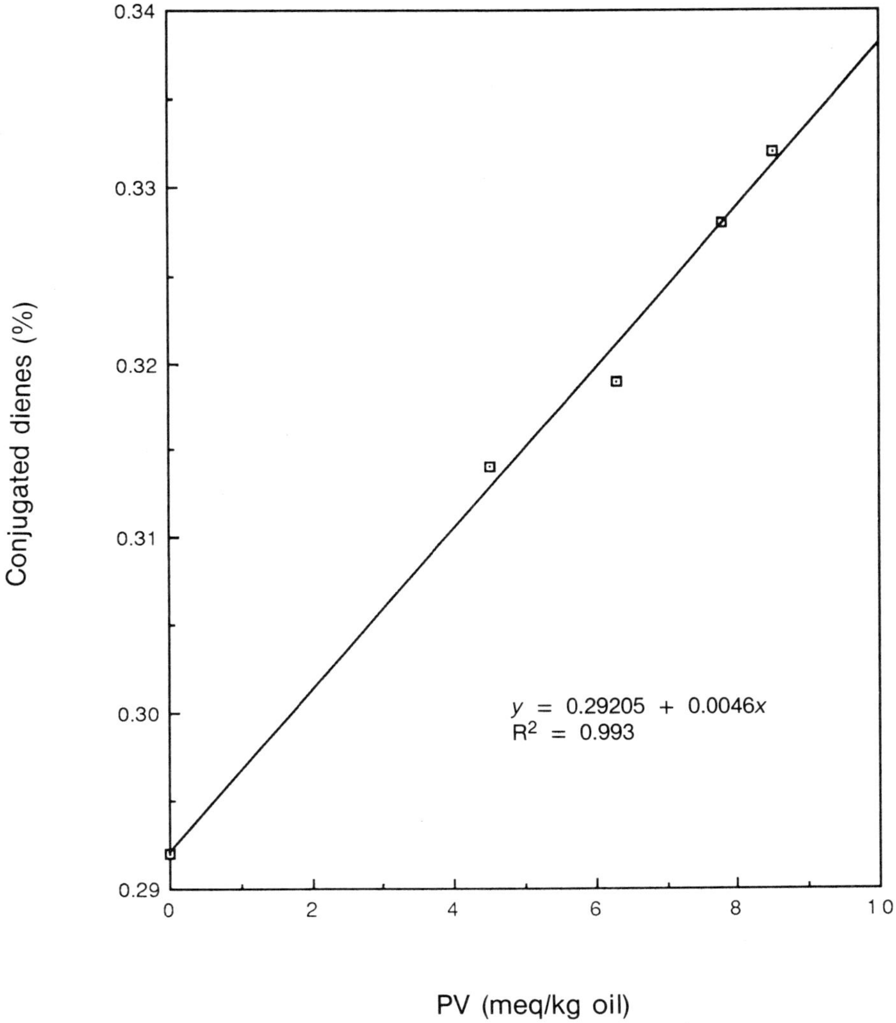


**Figure 8.6.** Relationship between peroxide value and absorbance at 233 nm of stored oil.

## Headspace Oxygen Analysis

Several analytical methods have been reported to determine the oxidative stability of oils by measuring the oxygen reacted with oils. The common oxidative stability determination methods in foods are gas chromatographic headspace oxygen determination, active oxygen method (AOM), oxygen bomb method, and polarographic

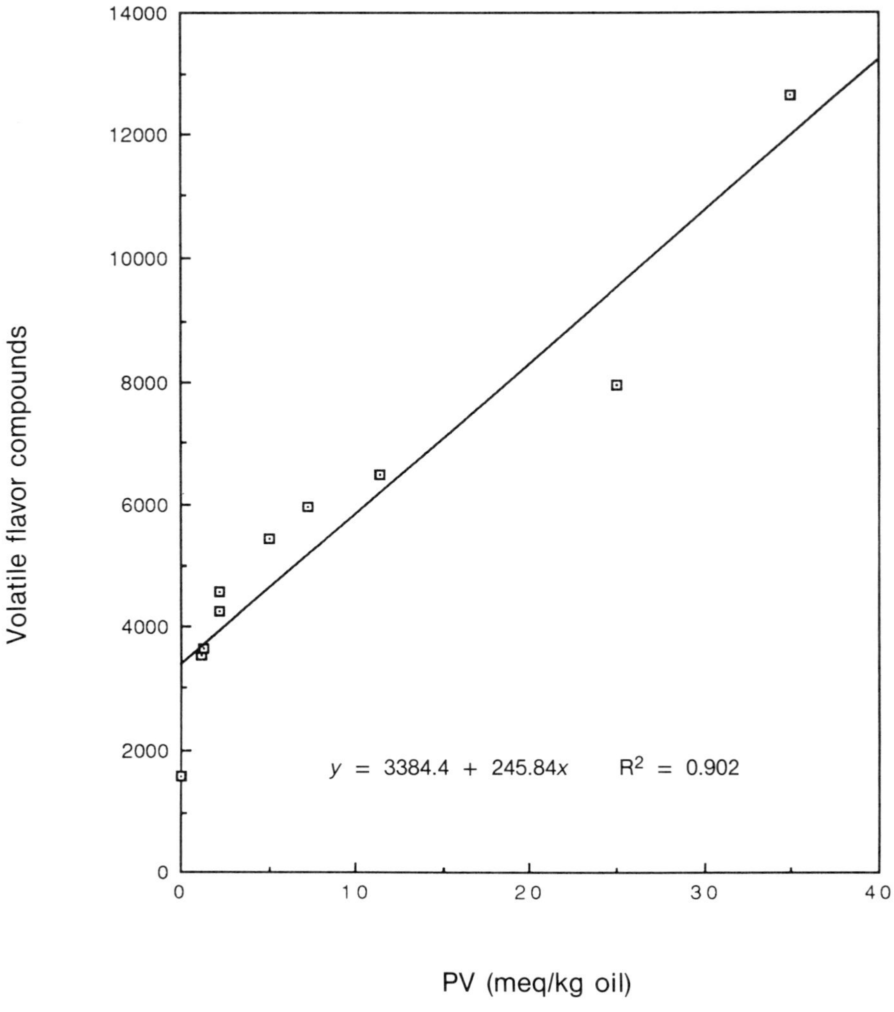


**Figure 8.7.** Relationship between peroxide value and volatile flavor compounds of stored oil.

dissolved oxygen determination method. Gas chromatographic headspace oxygen analysis is the simplest, most rapid, reproducible, and sensitive method to determine the oxidative stability of oils. This method can be used independently or to complement the Schaal Oven Method and AOM Factors affecting oxidative stability are temperature, pressure, and sample size and surface area.

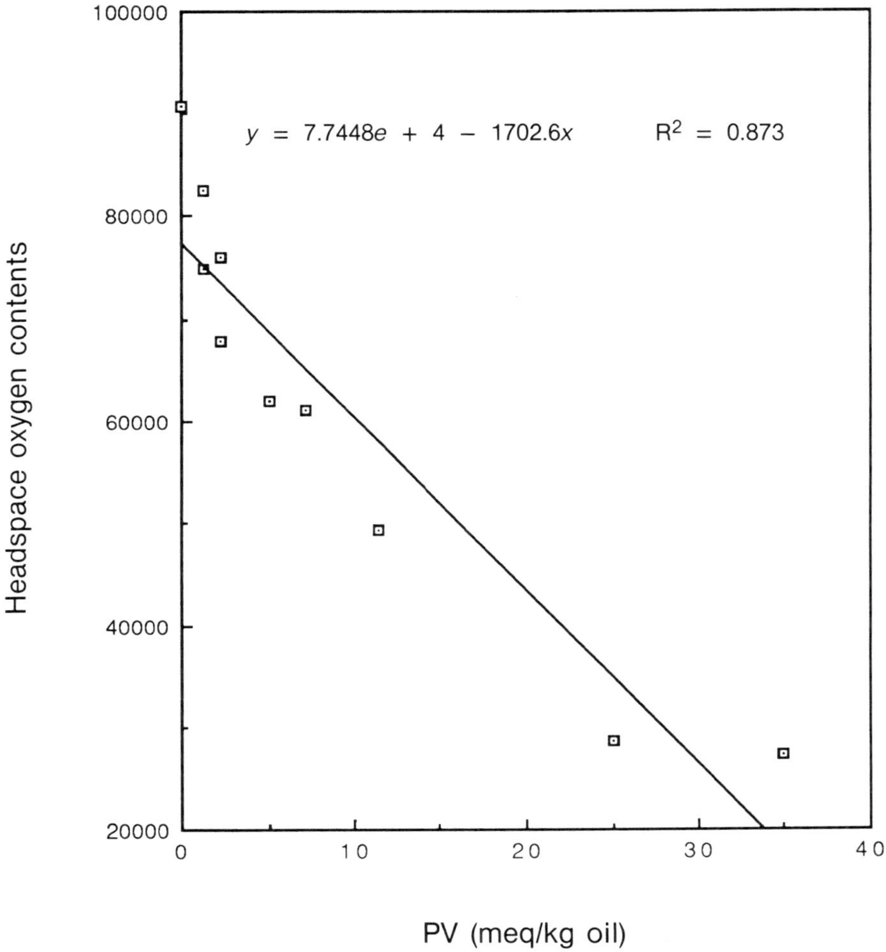


**Figure 8.8.** Relationship between peroxide value and headspace oxygen content of stored oil.

### *Headspace Oxygen Determination by Gas Chromatography*

To study the oxidative stability of oils or fats by gas chromatographic headspace oxygen analysis, 15 g oil is transferred to a 30-mL serum bottle (22–24). The bottles are sealed with Teflon™-coated rubber septa and aluminum caps and are stored in a forced-draft air-oven and evaluated for stability during storage. Headspace oxygen is analyzed using a thermal conductivity detector and a stainless steel column

**TABLE 8.2**
Relationship Between Peroxide Value and Sensory Score

| Aging Temperature (°C) | Sample | $R^2$ Values |
|---|---|---|
| 38 | Beef tallow | 0.93 |
| | Lard | 0.94 |
| | Hydrogenated soybean oil | 0.80 |
| | High-oleic safflower oil | 0.95 |
| | Corn oil | 0.84 |
| 60 | Beef tallow | 0.99 |
| | Lard | 0.90 |
| | High-oleic safflower oil | 0.97 |
| | Corn oil | 0.96 |

(183 × 0.32 cm) packed with 80/100 mesh Molecular Sieve 13X. Gas molecules are absorbed primarily because their molecular diameter matches the pore diameter of the molecular sieve. The thermal conductivities of helium, oxygen, and nitrogen at 500°K are 222, 41, and 38 (mW/mK), respectively (25). The thermal conductivity difference between oxygen and helium is greater than that between oxygen and nitrogen. Therefore, helium is recommended as the carrier gas to determine oxygen in the headspace of the sample bottle.

As the detector temperature is increased, the differences of thermal conductivity between oxygen and helium increases (25). High detector temperatures increase the detector response to oxygen. Forty microliters of headspace gas for a helium carrier or 1.0 mL of headspace for a nitrogen carrier has been used for the determination of oxygen in the headspace of an oil bottle (18–24,26,27). Temperatures of the injection port, oven, and detector were 200, 40, and 250°C, respectively. The oxygen peak in the headspace has been determined using an electronic integrator (26,27). The coefficient of variation for the determination of headspace oxygen content by gas chromatography has been calculated at 0.39% (Table 8.1).

Typical headspace oxygen depletion of an oil during storage is shown in Figure 8.9 (28). The headspace oxygen can be expressed as either the contents of the remaining oxygen in the headspace by concentration percentage (%) or as the remaining or reacted headspace oxygen in µmoles/mL. The conversion of headspace oxygen percent to µmoles/mL is possible by using the following relationships (26); 1 mL air contains 0.20946 mL $O_2$ (29) and 22,400 mL $O_2$ is equal to $10^6$ µmoles according to Avogadro's Law. Therefore, 0.20946 mL of $O_2$ is equal to 9.35 µmoles $O_2$. The area for gas chromatography peak of 9.35 µmoles of oxygen was measured in electronic counts by injecting 1 mL air into the gas chromatograph, then calculating the electronic counts of 1 µmole $O_2$. The peak area of the oxygen in 1 mL of headspace gas of the sample was measured in electronic counts and then the electronic counts were converted into µmoles $O_2$: µmoles of $O_2$ in 1 mL of headspace gas = electronic counts of $O_2$ in 1 mL of headspace gas × 1 µmole $O_2$/electronic counts of 1 µmole of $O_2$.

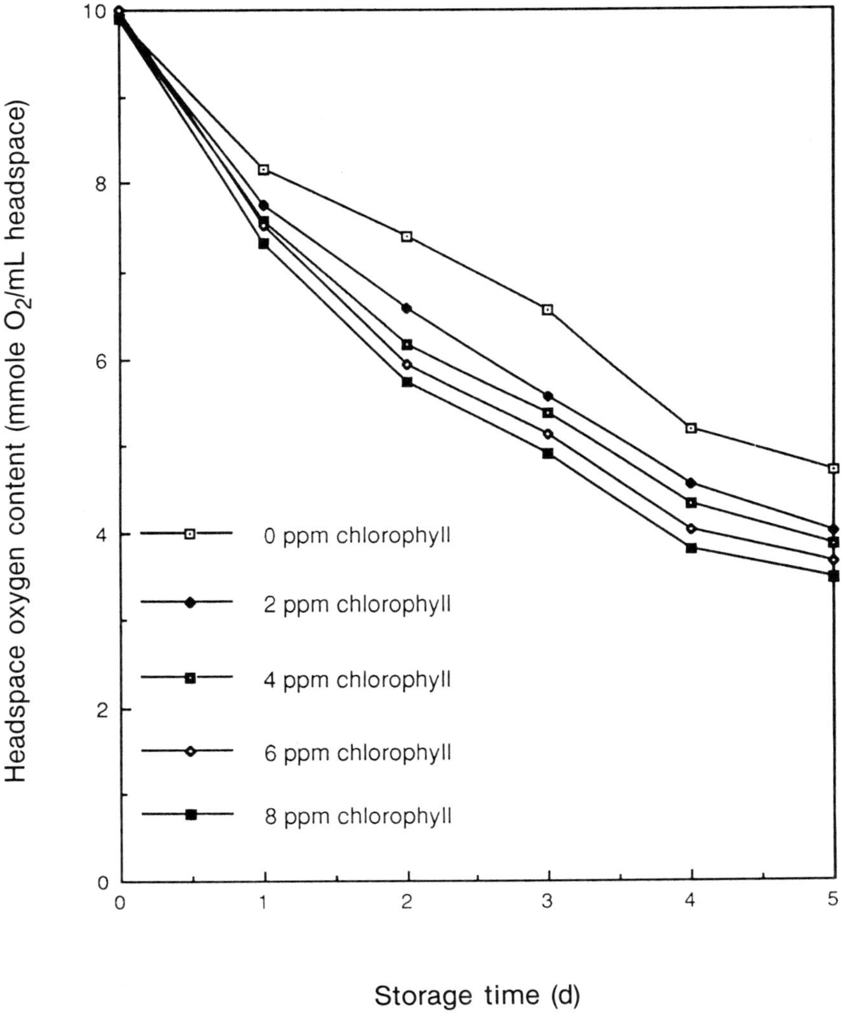


**Figure 8.9.** Depletion of headspace oxygen of stored soybean oil with various chlorophyll concentrations.

### *Relationship Between PV and Headspace Oxygen*

The PV of an oil is the quantity of hydroperoxide oxygen present in the sample, expressed in meq. of active oxygen per kg of the sample (30). Therefore, the depleted headspace oxygen in an oil could be converted to the PV of an oil in the initial oxidation period. A correlation coefficient of 0.87 was calculated between PV and headspace oxygen depletion, indicating a good relationship (Figure 8.8).

## References


1. Labuza, T.P. 1971. *CRC Crit. Rev. Food Technol. 2*:355.
2. Uri, N. In *Autoxidation and Antioxidants,* Vol. 1, Interscience Publishers, New York, 1961, p. 55.
3. Lundberg, W.O. In *Lipids and Their Oxidation,* AVI Publishing Co., Westport, CT.,1962.
4. Scott, G. *Atmospheric Oxidation and Antioxidants,* Elsevier Publishing Co., New York, 1965.
5. Emanuel, N.M., and Lyaskous-Kaya, Y.U. *The Inhibition of Fat Oxidation Processes,* Pergamon Press, New York. 1967.
6. Bradley, D.G., and Min., D.B. 1992. *Critical Rev. Food Sci. and Nutr. 31*:211.
7. Dugan, L.R. Jr. In *Principles of Food Science, Part 1: Food Chemistry,* edited by O.R. Fennema, Marcel Dekker Inc., New York, NY and Basel, Switzerland, 1976, pp. 139–203.
8. Madge, E.P. 1990. *The Relationship Between Soybean Oil Oxidation and the Decompositions of α–, β– oρ δ–Tocopherols During Storage.* M.S. Thesis. The Ohio State University.
9. Lea, C.H. In *Lipids and Their Oxidation Products,* edited by H.W. Schultz, E.A. Day, and R.O. Sinnhuber, AVI Publishing Co., Westport, CT, 1962, pp. 3–28.
10. Knight, H.B., Coleman, J.E., and Swern, D. 1951. *J. Am. Oil Chem. Soc. 28*:498.
11. Lea, C.H. 1931. *Proc. Royal. Soc. London 108B*:175.
12. Wheeler, D.H. 1932. *Oil Soap 9*:89.
13. *Official and Tentative Methods of the American Oil Chemists' Society, Vol. 1,* 4th Edition, American Oil Chemists' Society, Champaign, IL, 1989.
14. Daniel, S. *Bailey's Industrial Oil and Fat Products, Vol. 1,* John Wiley and Sons, 1979, p. 142.
15. Mehlenbacher, V.C. *The Analysis of Fats and Oils,* Garrard Press, Champaign, IL, 1960.
16. Gray, J.I. 1978. *J. Am. Oil Chem. Soc. 55*:539.
17. Fiedler, U. 1974. *J. Am. Oil Chem. Soc. 51*:101.
18. Jung, M.Y. 1989. *Effects of Carotenoids and Tocopherols on the Chlorophyll Sensitized Photooxidation of Soybean Oil.* Ph.D. Dissertation. The Ohio State University.
19. Min, D.B., and Schweizer. D.Q. 1983. *J. Am. Oil Chem. Soc. 60*:1662.
20. Min, D.B. *Food Lipids* (The Ohio State University). 1991.
21. Fioriti, J.A., Kanuk, M.J., and Sims, R.J. 1974. *J. Am. Oil Chem. Soc. 51*:219.
22. Jung, M.Y., Yoon, S.H., and Min., D.B. 1989. *J. Am. Oil Chem. Soc. 66*:118.
23. Yoon, S.H., Jung, M.Y., and Min., D.B. 1988. *J. Am. Oil Chem. Soc. 65*:1652.
24. Fakourelis, N., Lee, E.C., and Min., D.B. 1987. *J. Food Sci. 52*:234.
25. Lide, D.R. *CRC Handbook of Chemistry and Physics.* 72nd Edition 1991–1992. 6–162
26. Behroze, S.M., and Min, D.B. 1988. *J. Food Sci. 53*:1896.
27. Fakourelis, N., Lee, E.C., and Min, D.B. 1987. *J. Food Sci. 52*:234.
28. Lee, E.C. 1987. *Effects of Chlorophyll, β-Carotene and Tocopherol on the Singlet Oxygen Oxidation of Soybean Oil.* Ph.D. Dissertation. The Ohio State University.
29. Parker, S.P. (Ed.) *Encyclopedia of Science and Technology, Vol. 1,* 5th Edition, McGraw-Hill Book Co., New York, 1982.
30. Gunstone. F.D., Harwood, J.L., and Padley, F.F. *The Lipid Handbook.* Chapman and Hall. 1986. p. 263.

Chapter 9

# Conjugated Diene, Anisidine Value, and Carbonyl Value Analyses

**Pamela J. White**

Food Science and Human Nutrition Department, Iowa State University, Ames, Iowa 50011-1060, USA.

## Introduction

Many objective instrumental and chemical methods have been proposed over the years to evaluate the quality and stability of fats and oils. Generally, the methods are based on measuring one or more of the primary or secondary oxidation products, or a physical or chemical property resulting from such chemical changes. In some circumstances a particular method might be more useful than another. The usefulness of each procedure depends upon a number of factors, but the most important feature of a method should be how well data provided by the method correlates with the sensory properties of the fat or oil.

The purpose of this chapter is to evaluate methods for measuring three values commonly used for objectively measuring oil stability, conjugated diene, anisidine, and carbonyl values. Other chapters in this book evaluate additional instrumental, chemical, and sensory procedures.

## Conjugated Diene Value

### *Principles*

One of the first steps in the oxidation of linoleate (18:2) or higher polyunsaturated fatty acids (PUFA) in an oil is a shift in the position of the double bonds. The shift occurs as one hydrogen is lost from the methylene group positioned between two double bonds existing in the original 1,4-pentadiene configuration. A pentadienyl radical intermediate is produced that can shift to form one of two possible conjugated diene (CD) structures. These intermediate components react with molecular oxygen to form conjugated hydroperoxides. Early on, Farmer et al. (1) reported that the extent of double bond displacement correlated with the degree of peroxidation occurring in an unsaturated oil. A general outline of the autoxidation of 18:2 is presented in Figure 9.1. Double bonds can also shift to form CD during the hydrogenation or deodorization of an oil if the temperature rises above 245°C. Regardless of the route to formation, this conjugated structure absorbs strongly at a wavelength of 232–234 nm. The CD value is expressed as a percentage of conjugated dienoic

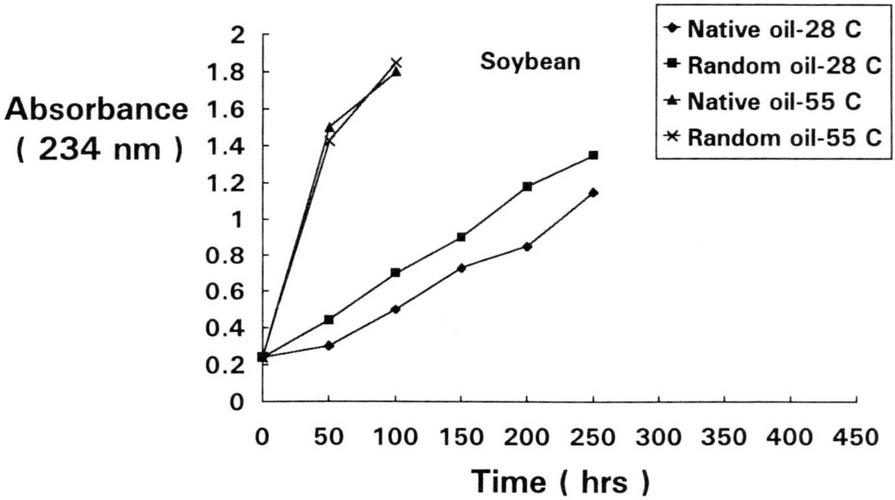


**Figure 9.1.** General reaction of the autoxidation of linoleate.

acid in the oil and is an indication of initial or primary oxidation products. Some researchers have used absorption at 268 nm to measure conjugated trienes as well as ethylenic diketones and conjugated ketodienes and dienals arising as secondary oxidation products (2).

The greater the amount of PUFA in an oil, the greater potential rise in CD. Moreover, the magnitude of change in CD cannot be related to the degree of oxidation, especially when comparisons are made among oils of different plant species or of different fatty acid profiles. Nonetheless, the CD value can be used as a relative measurement of oxidation in a particular oil of known fatty acid composition.

### *Procedures*

The American Oil Chemists' Society Method Ti 1a-64 is the most widely used method for measuring CD in an oil (3), using ultraviolet spectrophotometry to determine the diene conjugation of unsaturated linkages present. The procedure is applicable to dehydrated castor oil, dehydrated castor fatty acids, and their methyl or ethyl esters; because of the very high level of unsaturation in castor oil and its products, fairly high CD values result. This procedure can be used successfully for edible vegetable oils with high PUFA levels, resulting in much lower CD values than with castor oil.

To conduct the analysis, weigh 100 mg oil into a 100-mL volumetric flask and add 75 mL of purified isooctane. The flask is rotated and warmed to dissolve the sample completely. After cooling to room temperature, isooctane is added to volume with thorough mixing. Dilute the solution to give a final concentration of about 0.5 g/L for most edible oils. Each dilution step should not be more than 10:1. For castor oil, a final concentration of about 0.1 g/L is used. The absorbance of the solution at 233 nm is measured using a matching cell containing only purified solvent for the blank. The average of several readings is taken to obtain a mean value. The observed absorbency readings should lie between 0.2 and 0.8 to give an accurate measure of the value. If the reading is outside this range, then the dilution or the weight of the sample must be changed. The AOCS procedure should be consulted for notes about the apparatus and recommendations for running the tests.

The CD value is reported as a percentage of conjugated dienoic acid and is calculated from the formula:

$$\text{Conjugated dienoic acid, \%} = 0.84\,(A_s/bc - K_o),$$

Where:

$K_o$ = Absorptivity by acid or ester groups,
= 0.07 for esters,
= 0.03 for acids,
$A_s$ = observed absorbance at 233 nm,
$b$ = cell length (cm), and
$c$ = concentration of sample in g/L of the final dilution used for the absorption measurement.

The accuracy of the procedure is good, with the following guidelines given for precision. When the same operator and apparatus are used, repeatability results should not differ from the mean by more than ±0.39%. When different operators and instruments are used, reproducibility results should not differ from the mean by more than ±1.26%.

### Data Interpretation

Generally, when measuring oxidized vegetable oils, the CD values will range from 0–6%, depending upon the source of the oil. Obviously, the greater the amount of 18:2 and the higher the amount of unsaturated fatty acids in the oil, the greater the potential rise in the CD value. The CD accumulate to a certain percentage in any one oil and then plateau as the formed dienes are involved in additional oxidative reactions causing the dienes to break down. These reactions include scission to produce short-chain carbonyl compounds. Privett and Blank (4) suggested that the plateau of the UV curve always occurred just before the end of the induction period. The amount of CD accumulation increases with time and increased temperature. The CD value is particularly useful in measuring changes in oils during deep fat frying.

### Examples and Correlations with Other Tests

In 1975, St. Angelo et al. (5) reported a study of the oxidation of peanut butter stored at room temperature in the dark for up to 12 weeks. They measured the per-

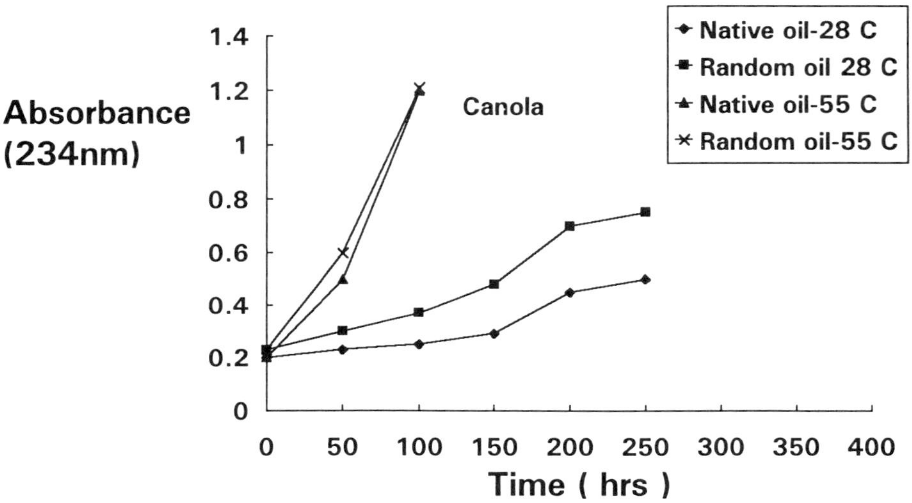


**Figure 9.2.** Lipid oxidation, as measured by absorbance at 234 nm, of native and chemically randomized soybean oils, Tautorus and McCurdy (6).

oxide values (PV) and CD values of peanut butter treated with several different additives. The correlation coefficient between the two measured values was 0.98 with a standard error estimate of ±2.29.

Tautorus and McCurdy (6) measured the oxidative stabilities of five native and chemically randomized oils stored at 28 and 55°C by using absorbance at 234 nm in hexane as an indication of CD content. Oxidation was carried out without stirring on 0.5 g quantities of oil in 15 × 45 mm vials in the dark, with measurements taken every 24 or 12 hr for oils stored at 28 or 55°C, respectively. The oils used included canola, corn, linseed, soybean, and sunflower oils. In every instance, the rate of increase of CD values as measured in both the native and the randomized oils was higher when stored at 55°C than at 28°C. Moreover, the highly unsaturated oils had a higher rate of accumulation of CD than did the other oils, with the highly unsaturated linseed and the more saturated corn and canola oils being the extremes. Figures 9.2 and 9.3 show the increase in CD values at 55 and 28°C, respectively, for native and chemically randomized soybean and canola oils (6).

Results reported in Figures 9.2 and 9.3 also demonstrate the effect of temperature on statistical significance. Differences between the random and chemically modified oils were more obvious at 28 than at 55°C. In fact, statistical differences of the slopes of these lines at 28°C were significant at $P < 0.05$ for every oil tested, whereas no significant differences were noted at 55°C between the two oil treatments for corn, linseed, sunflower, and soybean oils. Statistical analyses were not performed on the curves for canola oil at 55°C because no straight line portion with sufficient data points was observed.

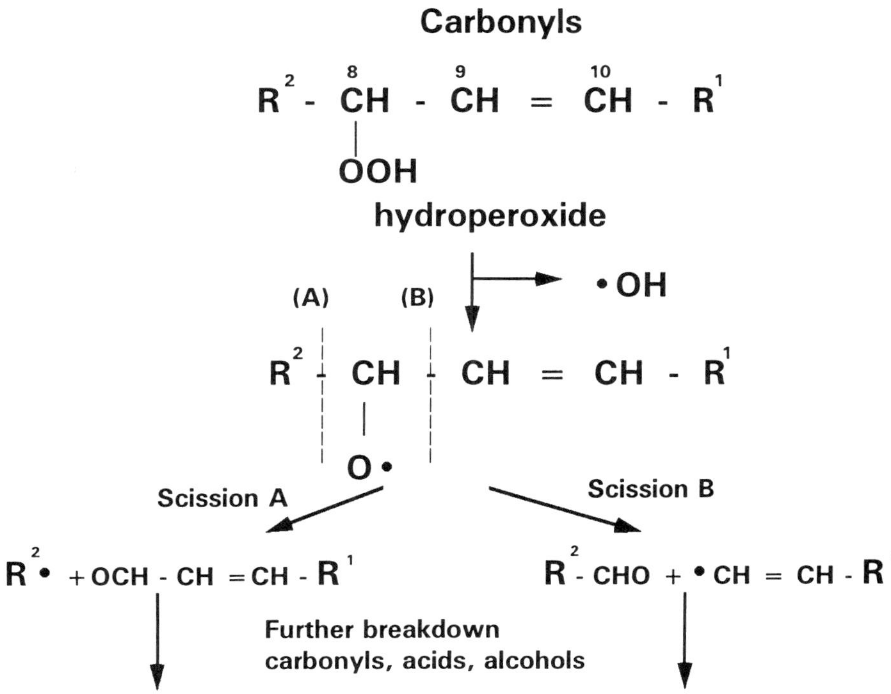


**Figure 9.3.** Lipid oxidation, as measured by absorbance at 234 nm, of native and chemically randomized canola oils. Tautorus and McCurdy (6).

White and Miller (7) compared stabilities of four soybean oils with different fatty acid compositions by measuring CD values during storage at 28 and 60°C (Tables 9.1 and 9.2). In both tables, the oils are listed in order of their 18:3 contents and in approximate order of their PUFA values. Oil types A5 and A6 had nearly identical PUFA values. The initial 18:3 content of the oils was 3.7, 6.3, 7.2, and 9.1% for A5, A6, Pella, and BSR 101 oils, respectively. The differences among the oil types were again greater when oils were stored at 28°C than at the higher tempera-

**TABLE 9.1**
Percentages of Conjugated Dienoic Acid[a-d] in Oils During Storage at 28°C[e]

| | Conjugated Dienoic Acid (%) at Day | | | | | | |
|---|---|---|---|---|---|---|---|
| Soybean Oil Type | 0 | 6 | 10 | 18 | 27 | 46 | 67 |
| Low Linolenic (A5) | $0.17^{b}$ | $0.14^{b}$ | $0.15^{a,b}$ | $0.25^{c}$ | $0.31^{c}$ | $0.36^{d}$ | $0.49^{d}$ |
| High stearic (A6) | $0.24^{a}$ | $0.20^{a}$ | $0.18^{a}$ | $0.28^{b}$ | $0.33^{b,c}$ | $0.46^{c}$ | $0.57^{c}$ |
| Pella | $0.14^{b}$ | $0.12^{c}$ | $0.13^{b}$ | $0.27^{b,c}$ | $0.36^{b}$ | $0.55^{b}$ | $0.80^{b}$ |
| BSR 101 | $0.16^{b}$ | $0.12^{c}$ | $0.13^{b}$ | $0.34^{a}$ | $0.48^{a}$ | $0.76^{a}$ | $1.04^{a}$ |

[a-d]Values in the same column with different superscript letters are significantly different ($P < 0.05$).
[e]White and Miller (7).

**TABLE 9.2**
Percentages of Conjugated Dienoic Acid[a-b] in Oils During Storage at 60°C[c]

| Soybean Oil Type | Conjugated Dienoic Acid (%) at Day 0 | 2 | 4 | 6 | 8 |
|---|---|---|---|---|---|
| Low linolenic (A5) | 0.19[a] | 0.23[b] | 0.31[a] | 0.49[a] | 0.69[a] |
| High stearic (A6) | 0.24[a] | 0.31[a] | 0.43[a] | 0.53[a] | 0.66[a] |
| Pella | 0.20[a] | 0.31[a] | 0.39[a] | NA | 0.74[a] |
| BSR 101 | 0.22[a] | 0.33[a] | 0.43[a] | 0.61[a] | 0.90[a] |

[a-b]Values in the same column with different superscript letters are significantly different ($P < 0.05$).
NA = not analyzed.
[c]White and Miller (7).

ture. In fact, by day 67 (the last day of storage), all oils were significantly different from one another when stored at 28°C, whereas no significant differences were noted among the oils at day 8 (the last day of storage) when they were stored at 60°C.

Miller and White (8) used CD measurements to determine differences among some of these same soybean oils with different fatty acid compositions during heating of the oils under deep fat frying conditions. In this test, Hardin oil with 6.9% 18:3 was substituted for Pella oil. Significant differences were noted among all oils by 30 hr of heating at 180°C (Table 9.3). Other research papers have also reported significant differences in CD values among oils or oil treatments during heating (9,10).

In contrast to the significant results just used as examples, we also have used the CD measurements in tests in which statistically significant results did not occur (11). The oxidative stabilities of one canola and six soybean oils with various fatty acid compositions were compared during storage at 60°C for 15 d. No significant differences were noted among CD values of the soybean oils after storage began, even though some significant differences were found in other conducted tests, including PV and sensory tests measuring flavor quality and flavor intensity of the oils. Evaluations among the tests and fatty acid compositions of the oils showed poor correlation of the CD values with most other values. The CD values did not correlate well with either the initial 18:3 content of the oils ($r = 0.63$, $P = 0.13$) or with the total initial PUFA value ($r = 0.13$, $P = 0.79$). Likely, the relatively warm temperature of 60°C used in this test narrowed any differences that might have been

**TABLE 9.3**
Percentages of Conjugated Dienoic Acid[a-d] in Oils Heated for Different Lengths of Time[e]

| Soybean Oil Type | Heating Time (hr) 0 | 10 | 20 | 30 | 40 |
|---|---|---|---|---|---|
| Low linolenic (A5) | 0.13[c] | 0.59[c] | 1.19[b] | 2.09[c] | 2.53[c] |
| High stearic (A6) | 0.16[c,d] | 0.52[c] | 0.98[b] | 1.58[d] | 2.01[d] |
| Hardin | 0.22[b] | 1.37[a] | 1.90[a] | 2.96[a] | 3.37[a] |
| BSR 101 | 0.62[a] | 1.07[b] | 1.58[a] | 2.75[b] | 3.13[b] |

[a-d]Values in the same column with different superscript letters are significantly different ($P < 0.05$).
[e]White and Miller (7).

found at lower oxidizing temperatures. Moreover, the complex mix of other ... acids present in the oils may have influenced the development of CD in the oils in an unpredictable fashion.

In a related study, these same oils were compared during intermittent heating and frying of bread cubes at 180°C (12). Although some tendencies were evident, few significant differences in CD values resulted. Once again, significant differences among the oils were measured with other tests, including PV of oils extracted from bread cubes fried in the oils and sensory tests of the fried cubes measuring flavor quality and flavor intensity of the oils. The CD values of oils heated for 40 hr were highly correlated with their initial PUFA contents ($r = 0.87$, $P = 0.01$), but the PV of oils extracted from stored cubes and the initial 18:3 contents of the oils were poorly correlated with the CD values of the heated oils ($r = -0.28$, $P = 0.54$; $r = -0.20$, $P = 0.67$, respectively). Again, the complex interaction of other fatty acids present in the oils may have influenced these results.

Yoshida et al. (13) studied the effects of microwave heating on the quality of several vegetable oils (linseed, soybean, corn, olive, and palm) with initial PUFA contents of 73.5, 63.1, 45,4, 7.3, and 9.4%, respectively. Their respective 18:3 levels were 57.7, 8.3, 1.9, 1.1, and 0.6%. Among other measurements, they determined CD values and absorbance at 268 nm during heating for up to 20 min. Before and after microwave heating, the CD values for olive and palm oils were significantly lower ($P < 0.05$) than those for the other oils, and the CD value for linseed was significantly higher than that for the other oils. Values for soybean and corn oils began to differentiate after 8 min in the microwave oven. Significant differences ($P < 0.05$) in absorbance at 268 nm did not occur among the oils until after 8 min of heating. Absorbance at 268 nm was not determined in olive and palm oils because of their low levels of 18:3. The differences in CD and absorbance at 268 nm among the oils closely followed the differences in their 18:3 levels and their initial PUFA contents.

Noor and Augustin (2) determined the effectiveness of commercial antioxidants (butylated hydroxyanisole and butylated hydroxytoluene) in improving the stability of banana chips fried at 190°C and then stored at 65°C for up to 50 d. They used several methods, including absorbance at 232 nm and 268 nm, to measure the oxidation of oils extracted from the chips during storage. They noted that changes in absorbance at 232 nm closely followed changes in PV, although they did not report correlation values. They also noted an increase in absorbance at 268 nm during storage time, but the pattern of increase differed somewhat from measures of PV and CD values, likely because of the different end-products being measured by these methods. All methods, however, showed the same general trend for protection of the chips fried in oils with the commercial antioxidants.

Fishwick and Swoboda (14) described The Conjugable Oxidation Product (COP) assay, a procedure for measuring the extent of oxidative deterioration of PUFA. The method is based on the conversion of fatty acid hydroperoxides and their derivatives, oxodienes and hydroxydienes, to conjugated fatty acids with strongly chromophoric triene and tetraene structures. The conversion is done in two steps, reduction with sodium borohydride followed by dehydration under acidic conditions. The chromophores produced are measured by absorbance in the UV range. They used methyl linoleate and methyl linolenate as test compounds for the procedure. Dienoic fatty acid oxidation products created a conjugated triene chromophore, whereas the more highly unsaturated fatty acids formed a conjugated tetraene chromophore. The sum of absorbance changes at 268 and 301 nm was used as a measure of the degree of oxidation. The oxygen uptake in the presence of methyl linoleate was plotted against both PV and absorbance of triene chromophore (268 nm), and the relations were linear.

The ratio of the absorbance change from 301 to 268 nm, termed the COP ratio, varied from zero for a pure triene to 2.8 for a pure tetraene. Thus, this ratio is a measure of the type of PUFA being oxidized and could help distinguish between the oxidation products derived from the dienoic and the polyenoic fatty acids. They also noted that the concentration of oxodiene component present in the original mixture may be estimated separately by a decrease in the 275 nm region during the reduction step.

Parr and Swoboda (15) applied the COP method to fresh and stored foods and fats. They concluded that the PV and the COP value had the same level of numerical significance.

### *Advantages and Disadvantages of the CD Method*

The CD method measures primary oxidation products occurring during the oxidation of fats and oils, and its value correlates fairly well with PV. The CD method offers several advantages over the PV method in that it is faster, simpler, requires no chemical reagents, does not depend upon a chemical reaction or color development, and can be conducted on small (about 100 mg) sample sizes. The CD method can also be used to measure changes in oils held at frying temperatures (about 180°C). Unlike peroxides, which are very unstable at high temperatures and cannot be used as a progressive measure of high-temperature oxidation, CD form and accumulate fairly rapidly in oils containing PUFA.

The CD method, however, cannot be applied to all studies of oil stability, because the magnitude of the value obtained depends mainly upon the fatty acid composition of the fat or oil. Therefore, CD values cannot be compared from one species of oil to another easily, especially when large differences in the beginning fatty acid compositions occur. Moreover, the CD procedure is useful in measuring changes only in oils containing substantial amounts of linoleate or more highly unsaturated fatty acids, such as most liquid vegetable oils.

Some solvents and foods interfere with absorption of CD at the desired wavelength, so the sample to be measured must be free from any contaminating materials. Finally, as previously mentioned, the CD method measures primary breakdown products occurring in the early stages of oxidation, and the accumulation of these products parallels the uptake of oxygen and the formation of peroxides in the early stages of oxidation (16). The CD values, however, plateau at a certain concentration, at which level the breakdown of formed CD presumably equals formation of new CD compounds. At this point in oxidation, no further changes can be measured by the CD procedure.

## Carbonyl Value

### *Principles*

When hydroperoxides such as those shown in Figure 9.1 decompose, they form a number of secondary oxidation products, for example carbonyls (aldehydes and ketones), lactones, ethers, alcohols, acids, furans, and hydrocarbons. Figure 9.4 gives a generalized scheme for the breakdown of hydroperoxides to their initial scission products, which then can undergo additional reactions to form other secondary oxidation products.

Carbonyls are considered the chief compounds responsible for oxidized flavors, so many methods have been proposed to specifically determine carbonyl com-

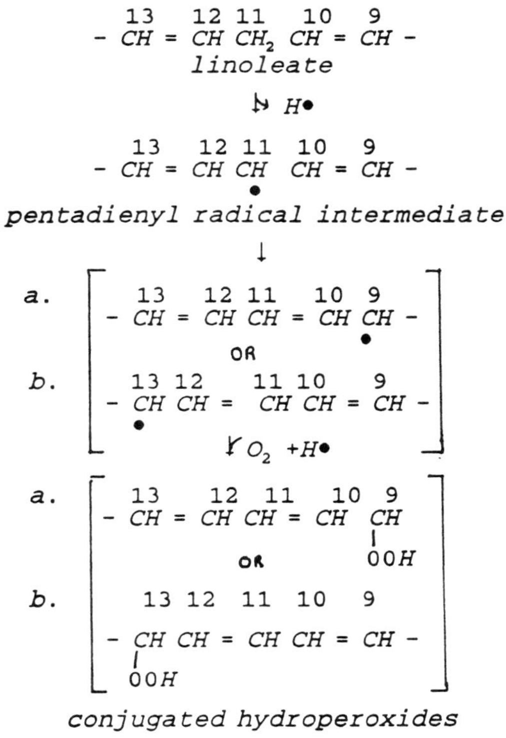


**Figure 9.4.** Generalized scheme for the breakdown of hydroperoxides to their initial scission products.

pounds for possible correlation with sensory evaluations. Most methods determine the total carbonyl content which includes free volatile carbonyls, free nonvolatile carbonyls, and carbonyls present in bound form (such as aldehydo-glycerides [17]). It is the volatile carbonyls, however, that contribute most to off-flavors in oxidized fats; thus, a variety of techniques have been developed to separate and measure these volatile components. Generally, vacuum or steam distillation is used to recover the volatiles, which then can be measured separately. The measurement of volatile components from oxidized oils is now commonly accomplished by means of gas-liquid chromatography (Chapter 6).

### *Procedures*

The most widely used measure of total carbonyl compounds is described by Henick et al. (18). A similar procedure for carbonyl value (CV) has been adopted as standard Method No. 2.4.22-73 of the Japanese Oil Chemists' Society (JOCS [19]). The procedure is based on the formation of 2,4-dinitrophenylhydrazones of the carbonyl compounds in the presence of a trichloroacetic acid catalyst. The resulting colored hydrazones can be measured spectrophotometrically in an alkaline solution.

A brief description of the measurement of total carbonyl compounds follows. All reagents must be carbonyl-free, and procedures for their purification have been described (18). The fat sample (0.05–0.08 g) is weighed into a 50-mL volumetric flask to which is added 5 mL of benzene, 3 mL of 4.3% trichloroacetic acid solution in benzene, and 5.0 mL of 0.05% dinitrophenylhydrazine in benzene. The solution is incubated for 30 min at 60°C and then cooled to approximately 25°C. To develop the color, 10 mL of 4% potassium hydroxide in ethanol is added, and the solution is diluted to 50 mL with absolute ethanol. After exactly 10 min, the absorbance is read against a reagent blank prepared in the same manner, but without the fat sample. The printed procedure should be consulted for precautions necessary in performing the analysis.

Henick et al. (18) reported maximum absorption of the 2,4-dinitrophenylhydrazones of saturated aldehydes at 432 nm and of unsaturated aldehydes at 458 nm. They described standardizations for the simultaneous determination of saturated and allenic carbonyl contents from the absorbencies at the convenient wavelengths of 430 and 460 nm as follows:

$$\text{Unsaturated} = \frac{3.861\, A_{s460} - 3.012\, A_{s430}}{0.854}$$

and

$$\text{Saturated} = 3.861\, A_{s460} - 2.170\ \text{Unsaturated}.$$

They also reported a maximum absorption at 440 nm of hydrazones formed from lard, and used this wavelength to test the reproducibility of the method.

Triplicate lard samples were measured in duplicate analyses. At a 95% confidence interval, the CV of the lard was 5.70 ± 0.164 μmol/g. When lard was extracted from shoestring potatoes (5 replicates), the CV of the fat was 10.71 ± 0.33 μmol/g. The JOCS standard method uses absorbance at 440 nm to measure total CV, which is expressed as μmol hexanal/g fat or as μmol/kg fat. In a well-refined oil, CV should be in the range of 0.5–2.0 μmol/g fat.

### Data Interpretation

The CV procedure has been criticized, because hydroperoxides decompose under the conditions of the test and thus possibly interfere with the quantitative determination of the carbonyls. Several modifications to this procedure have been suggested to overcome the problem. Mizuno and Chipault (20) reduced the hydroperoxides to noncarbonyl compounds by adding stannous chloride reagent before carbonyl determinations, thus eliminating the interference from hydroperoxides. However, Fioriti (21) suggested that reduction generates additional carbonyls and is time consuming. He carried out the reaction at 5°C to decrease the decomposition of hydroperoxides to carbonyls. The low-temperature reaction, however, required 20 hr in contrast to the 30 min at 60°C. No single solution is ideal.

### Examples and Correlations with Other Tests

The CV method has not been widely used in recent years probably because of inaccuracies in the procedure as just described, the lack of correlation with sensory evaluations, and the development of more sensitive and precise ways of measuring specific volatile carbonyls by means of gas-liquid chromatography. Some recent studies have used CV along with other measures of oil quality.

Yoshida et al. (13) compared the effects of microwave heating for up to 20 min on the chemical properties of five seed oils (linseed, soybean, corn, olive, and palm) by means of several procedures, including CV, CD, and anisidine value (AV). The general tendencies for CV were similar to those for AV, and except for linseed oil, significant differences ($P < 0.05$) did not show up until after heating for 12 min (Figure 9.5). The CD method showed significant differences in the oils after just 8 min of heating.

Chu (22) did a comparative study of analytical methods for evaluating oil quality during heating and frying by studying soybean oil alone, in blends with palm and sesame oil, and with the addition of antioxidants and metal chelators. Table 9.4 shows correlations between the values measured with six procedures and that of changes in dielectric constant (DC) obtained with the Food Oil Sensor (FOS [23]). These values were reported to be highly significant ($P < 0.001$). Both the CV and the AV gave relatively high correlations with the DC.

An FOS reading of 3.7 or higher (24,25) or of 4.0 or higher (26) has been suggested as the limit beyond which a used fat should be discarded. Chu (22) selected a FOS value of 4.0 to indicate the discard value for frying oils and related that read-

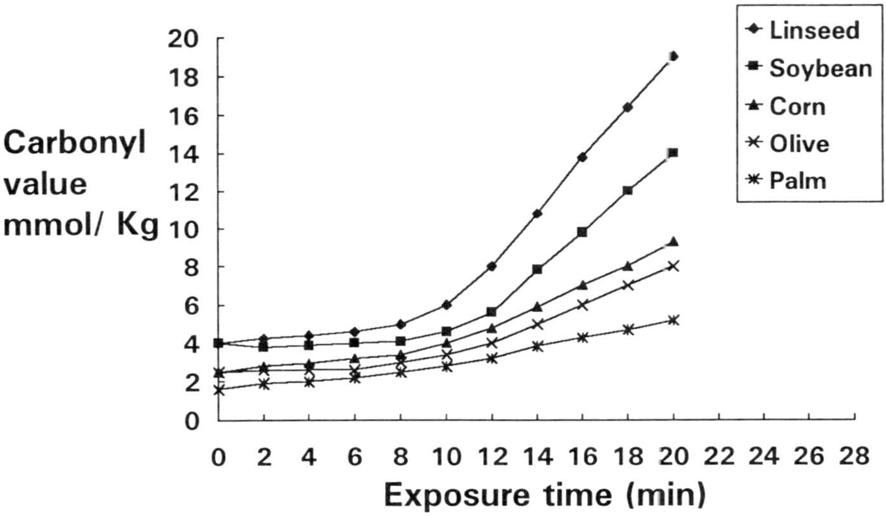


**Figure 9.5.** Effect of microwave exposure time on carbonyl values of selected seed oils, Yoshida et al. (13).

ing to values obtained with other tests. The FOS reading of 4.0 corresponded to the following values: total polar compounds, 22%; Fritest®, 0.2175; refractive index, 1.4750 $n_D$; iodine value, 126.2 g/100 g fat; AV, 154.6 µmol/g fat; CV, 42.2 µmol/g fat; and acid value, 2.7 mg KOH/g fat. Overall, Chu found no single procedure that produced reliable results in all instances; however, FOS, total polar compounds, refractive index, and iodine value were more reliable than the rest of the methods assessing oil stability.

Boki et al. (27) evaluated the effects of filtration bleaching on PV, thiobarbituric acid value (TBA), and CV of autooxidized soybean oil, by using 23 kinds of activated carbon. The effects on decreasing PV, TBA, and CV of the autooxidized soybean oils differed among the types of activated carbons, but the relations among decreases in PV, TBA, or CV and specific surface area, basicity, or pore volume

**TABLE 9.4**
Correlation Between Changes in Dielectric Constant and Other Analyses of the Combined Data of Soybean Oil with Different Additives During Heating and Frying[a]

| Analysis | Number of Samples | Correlation Coefficient |
|---|---|---|
| Fritest® | 80 | 0.89 |
| Total polar compounds | 100 | 0.87 |
| Refractive index | 100 | 0.92 |
| Iodine value | 100 | 0.80 |
| Anisidine value | 100 | 0.82 |
| Carbonyl value | 100 | 0.87 |

[a]Chu (22).

with pore radii 12.5–110 Å were significant ($P < 0.05$). The authors suggested that hydroperoxides, aldehydes, and ketones were adsorbed on the acid sites distributed over the surface or within the pores of the activated carbon, while the autooxidized soybean oil flowed through the packed column containing the carbon. The correlation coefficient between PV and acidity was greater than those between PV and specific surface area or pore volume. This finding suggested that the polarity of hydroperoxides enhanced their adsorption on more acidic columns. In contrast, the correlation coefficients between TBA or CV and these properties were reversed, indicating that adsorption of aldehydes and ketones (compounds measured by TBA and CV) was more affected by greater surface area or greater pore size than by acidity. Thus, all methods, including CV, gave useful information to the researchers.

### *Advantages and Disadvantages of the CV Procedure*

The CV procedure is simple and sensitive, and the theory behind measuring the carbonyls formed during oil oxidation is a good one (i.e., to measure the actual flavor compounds formed). Carbonyl value does, however, have many disadvantages. For example, the solvents must be purified to eliminate interference from existing carbonyls, and sample extracts should be run within 1 hr to obtain an accurate answer. Most importantly, the hydroperoxides decompose under the experimental conditions, causing the formation of additional carbonyls not originally present in the oil being tested. Some solutions to these problems have been suggested (20,21), but even so, the fact that the CV procedure measures total carbonyls, including those that are nonvolatile, probably decreases the ability of this procedure to correlate well with sensory tests.

## Anisidine Value

### *Principles*

The formation of secondary oxidation products as shown in Figure 9.4 and as described in the previous section on CV is also used as the basis for AV. Many of the compounds considered responsible for the off-flavor development in oils are actually aldehydes, so the AV procedure was developed to use the reaction of α- and β-aldehydes (primarily 2-alkenals) with *p*-anisidine reagent. When the reaction takes place in an acetic acid solution, the resulting Schiff base compounds create a yellowish color measurable at a light absorbance of 350 nm. The molar absorbance increases if the aldehyde contains a double bond conjugated to the carbonyl double bond; thus, the AV is mainly a measure of 2-alkenals.

### *Procedures*

The AV procedure is based upon a test initially developed by Holm et al. (28) involving the reaction of aldehydes with benzidine acetate. The reaction does not require a high temperature or strong acid, so the creation of artifacts from hydroperoxide breakdown, a serious problem in conducting the CV procedure, is minimized. Because benzidine is carcinogenic, an alternative procedure was proposed in which *p*-anisidine (*p*-methoxyaniline) is substituted for benzidine (29). The *p*-anisidine procedure has since been adopted as standard method 2.504 of the International Union of Pure and Applied Chemistry (IUPAC [30]).

Holm (29) suggested that it would be practical to have an expression for "the sum total of peroxides and secondary oxidation products" in an oxidizing oil. Because peroxides are decomposed to give secondary oxidation products that are anisidine-reactive, it was possible to relate AV to PV. Holm showed that an increase by one PV unit corresponds to an increase of about 2 AV units and suggested the term "oxidation value" (OV) to describe the degree of oxidation for an oil, in which OV = AV + 2 (PV). This calculation also has been termed the "Totox" value (31). Both the Totox value and AV are used more extensively in Europe than in the United States.

A brief outline of the IUPAC *p*-anisidine procedure (30) is described. The oil sample (0.5–4.0 g) is diluted to 25 mL with isooctane in a volumetric flask. The absorbance of the fat solution is read at 350 nm. Five milliliters of the fat solution is mixed with 1 mL of 0.25% p-anisidine in glacial acetic acid and, after 10 min, the absorbance at 350 nm is read against a reagent blank. The AV is given by the formula

$$AV = 25 \times (1.2A_s - A_b)/m,$$

Where:

$A_s$ = Absorbance of the fat solution after reaction with the *p*-anisidine reagent,
$A_b$ = Absorbance of the fat solution alone, and
$m$ = Mass of the sample (g).

The AV is defined as 100 multiplied by the absorbance measured in a 1-cm cell of a solution resulting from the reaction of 1 g of fat with a 100-mL mixture of solvent and reagent. In most instances, *n*-hexane can be substituted for isooctane. Highly oxidized oils, however, will not dissolve completely in *n*-hexane, so these oils will require isooctane as the solvent. The printed method should be consulted for notes about running the procedure.

The AV procedure was studied for its reproducibility and accuracy in analyzing oils (29). Beers law held for a solution of 1–16 g of oil containing 0.1–10 µmol 2,4-dienals/100 mL oil. Collaborative analyses among seven laboratories analyzing five different types of oils showed intralaboratory variance of 0.11 and interlaboratory variance of 0.17. List et al. (32) obtained a standard deviation of 0.09 AV units by analyzing a number of soybean oils on different days.

## Data Interpretation

Well-refined oils generally have AV between 1.0 and 10.0 mmol/kg fat (17). In oils with high PUFA levels, such as sunflower oil, there is a high potential for the formation of 2-alkenals. Thus, these oils may have AV higher than 10.0 mmol/kg even when fresh. Snyder et al. (33) reported that high-quality crude soybean oils generally had a Totox value of less than 3 and that crude sunflower oils had a Totox value of less than 5.

### *Examples and Correlations with Other Tests*

List et al. (32) evaluated soybean salad oils (crude, partly processed, and nonhydrogenated) from sound and damaged soybeans during storage at 60°C with and without light exposure. Table 9.5 shows the effect of 60°C storage on the quality and oxidation of nonhydrogenated soybean oil from sound beans, as measured by flavor scores, PV, AV, and OV or Totox values. The influence of fluorescent light exposure on the same type of soybean oil is shown in Table 9.6. A highly significant correlation (−0.68, $P < 0.01$) was found between the AV of salad oils from sound soybeans and their flavor scores (Table 9.7). Multiple correlations among flavor scores, AV, and PV of sound soybean oils yielded an even higher correlation of 0.81 ($P < 0.01$). Testing oils from damaged beans resulted in poor correlations among the same methods (Table 9.7). In fact, analysis of damaged oils before and after deodorization showed little, if any, reduction in AV, whereas deodorization of sound oils lowered the AV. The poor quality of the oil from damaged soybeans was substantiated by sensory evaluations. Henning (17) suggested that AV may be use-

**TABLE 9.5**
Effect of Accelerated Temperature Storage Test Upon Quality and Oxidation of Soybean Salad Oil[a]

| Test Conditions | Flavor Score | Peroxide Value | Anisidine Value | Oxidation Value |
|---|---|---|---|---|
| Control | 6.9 | 0.1 | 2.0 | 2.2 |
| 2 d, 60°C | 6.2 | 0.5 | 2.6 | 3.6 |
| 4 d, 60°C | 5.5 | 0.6 | 2.2 | 3.4 |
| 6 d, 60°C | 5.3 | 2.2 | 2.2 | 6.6 |
| 8 d, 60°C | 5.7 | 5.5 | 2.9 | 13.9 |
| 10 d, 60°C | 3.4 | 5.9 | 3.4 | 15.2 |

[a]Nonhydrogenated commercial salad oil. List et al. (32).

**TABLE 9.6**
Effect of Accelerated Light Storage Test Upon Quality and Oxidation of Soybean Salad Oil[a]

| Test Conditions | Flavor Score | Peroxide Value | Anisidine Value | Oxidation Value |
|---|---|---|---|---|
| Control | 6.9 | 0.1 | 2.0 | 2.2 |
| 2-hr light exposure | 6.0 | 1.0 | 2.2 | 4.2 |
| 4-hr light exposure | 6.3 | 1.3 | 2.2 | 4.8 |
| 8-hr light exposure | 4.5 | 1.6 | 2.0 | 5.2 |
| 16-hr light exposure | 4.9 | 2.6 | 2.0 | 7.2 |

[a]Nonhydrogenated commercial salad oil. List et al. (32).

**TABLE 9.7**
Correlation of Variables with Initial Flavor Scores[a]

| Variable | Correlation Coefficient (r) and Significance[b] | | |
|---|---|---|---|
| | Sound Oil[c] | Damaged Oil[d] | Pooled Data |
| Flavor and peroxide | −0.52* | 0.13 | −0.24 |
| Flavor and anisidine | −0.68** | −0.02 | −0.61** |
| Flavor, peroxide, & anisidine | 0.81** | 0.01 | 0.65** |
| Flavor and oxidation value | −0.80** | 0.13 | −0.65** |

[a]List et al. (32).
[b] *= Significant at 95% level; ** = Significant at 99% level.
[c]$n$ = 19.
[d]$n$ = 16.

ful for judging the degree of oxidative damage done to crude oils and for following the efficiency of the refining process. List et al. (32) reported that the AV as calculated into a Totox value provided a method for generally predicting the initial flavor scores of salad oil processed from undamaged soybeans.

Noor and Augustin (2) used AV along with several other measures of oxidation to evaluate the effectiveness of antioxidants on improving the stability of fried banana chips stored at 65°C. They were able to detect differences among treatments by using AV, but the progression of values was not as direct as with the PV and CD measurements (data described earlier in this chapter). Both AV and absorbance at 268 nm, which measure secondary oxidation products, showed the same general trends for oxidation of the treatments, but there was a difference in the quantity of compounds measured by these two methods. The difference could be attributed to the types of secondary compounds measured by each method, with AV being particularly sensitive to 2-alkenals and absorbance at 268 nm measuring ethylenic diketones, conjugated ketodienes, dienals, and trienes. No sensory evaluation was reported in the study.

Earlier in this chapter, a research study by Yoshida et al. (13) was discussed which examined the effects of microwave heating on the chemical properties of five seed oils. The relations among AV, CV, and CD values were also discussed. As described, AV and CV showed the same basic tendencies in oxidation of the oils. Carbonyl values, however, gave more differentiation than did AV. Figure 9.6 presents AV from that study, which can be compared to CV in Figure 9.5.

Anisidine value was shown to lack sensitivity in the research work of Schnepf et al. (34), in which menhaden/soybean oil blends were stored for up to 20 weeks at 22°C in the dark, either with or without oxygen in the headspace. Peroxide values increased slowly and consistently over time, reaching higher values when more menhaden oil was present or when air was present in the headspace. Anisidine values also increased with greater amounts of menhaden oil in the blend, but changed during the storage period only for the 100% menhaden oil stored with air in the headspace. Sensory evaluations of salad dressings made with the oil blends and stored up to 8 weeks showed differences in the salad dressings even when the PV

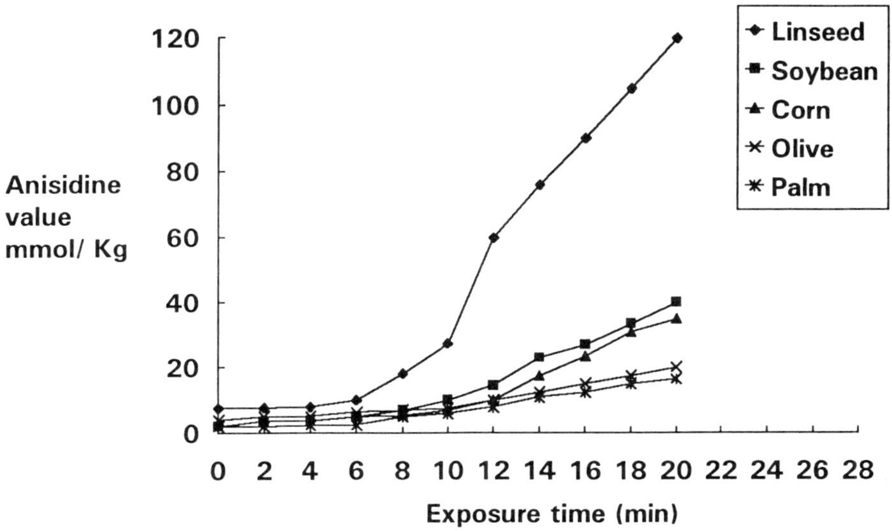


**Figure 9.6.** Effect of microwave exposure time on anisidine values of selected seed oils, Yoshida et al. (13).

for the stored oil blends was still low but, like PV, did change over time. The detection of rancidity by sensory panels usually is more sensitive than instrumental or chemical measures, but the fact that the salad dressings were stored in an emulsion form likely contributed to faster oxidation in this system than in the straight oil blends. Nonetheless, AV was of little use during the study in detecting oxidative development in the oil blends over time.

An analysis scheme for rapidly estimating the quality of an extracted crude oil while still in the seed was proposed by Snyder et al. (33). Factors of crude oil quality evaluated were free fatty acids, oxidative status (Totox value), color, and phosphatides (soybean) or wax (sunflower). Soybean and sunflower seeds subjected to extended storage at various moisture contents were sampled over time to yield 50 storage-damaged samples of each oilseed. Crude oil was extracted from the seeds of each sample and analyzed for the crude oil quality factors according to standard methods which were correlated with alternative instrumental and chemical procedures adapted to analyses of the seeds or of a quick extraction from the seeds. For example, hexanal content measured by headspace-gas chromatographic analysis of the ground full-fat meal was correlated with the Totox value of the crude oil. Other quick tests were proposed that correlated with color, and phosphatide or wax content.

In previous work, hexanal was shown to be one of the primary volatiles formed in oxidized soybean and sunflower oils (35). The best equations in the work of Snyder et al. (33) were obtained when the hexanal content determined by volatile headspace of the meal was correlated with the Totox value. Regression equations were Totox = 1.1 + 0.21 × hexanal (ppm) for soybeans (SD = 1.2, $r^2$ = 0.94) and Totox = 1.6 + 0.85 × hexanal (ppm) for sunflower seeds (SD = 1.3, $r^2$ = 0.72). Snyder et al. (33) did further analyses on soybean and sunflower seed samples collected from various sources during the 1985 and 1986 crop years. For soybeans, the Totox values estimated from hexanal concentration (mean = 4.4, SD = 5.1) were not consistently different (as determined by using analysis of variance, $P < 0.05$) from the Totox values from the analytical method (mean = 3.9, SD = 6.9). The values from the two methods had a correlation coefficient of $r = 0.83$. For sunflower seeds, the Totox value estimated from hexanal concentration in the ground meal (mean = 5.5, SD = 2.9) was not different from the analytical Totox (mean = 6.7, SD = 3.8) and the two methods were correlated ($r = 0.71$). In general, use of the headspace volatile-GC analysis to measure hexanal content on ground meal was a good predictor of the Totox value of the extracted crude oil.

Idris et al. (36) correlated sensory (flavor intensity and overall quality) and chemical (free fatty acid, PV, AV, and Totox value) analyses of palm oils of various qualities. Correlation coefficients between overall quality and free fatty acid content for red palm oils was $r = -0.80$, and for refined, bleached, and deodorized (RBD) palm oils it was $r = -0.69$. The significance level was not reported. Flavor intensity scores and free fatty acid values were correlated at $r = 0.61$ for RBD palm oils and at $r = 0.58$ for red palm oils. Correlation coefficients between overall quality and PV, AV, and Totox values of RBD palm oils were $r = -0.57$, $r = -0.54$, and $r = -0.50$, respectively ($P < 0.10$). Slightly better correlations of about −0.61 were obtained for red palm oils. Correlation values for flavor intensity scores in RBD and red palm oils and PV, AV, and Totox values were similar to values just listed. All correlations were positive, but most were significant only at $P < 0.10$. Thus, only general predictions of sensory quality of the oils could be made by any of the chemical tests, and the free fatty acid value was better at predicting the sensory properties than were PV, AV, or Totox.

### *Advantages and Disadvantages of the AV Method*

The AV procedure is a quick and simple one performed directly on a solution of fat or oil. The reaction occurs in fats and oils without heat or strong acid, so the presence of hydroperoxides does not interfere with the reaction. The procedure can be used to monitor changes during oil processing, especially when the history of the oil is known.

Anisidine values are not very sensitive and are not particularly suited for predicting off-flavors in oils. The procedure mainly measures 2-alkenals, which are not always the compounds causing off-flavors. It is probably most useful for routine testing of crude oils to determine treatment before processing rather than as a research tool. Moreover, the value obtained is comparable only within each oil type, because initial AV vary among oil sources.

With regard to the procedure itself, the reagents must be moisture free, because the presence of water leads to incomplete reactions. As with the CV procedure, all reagents must be carbonyl free to avoid interference from existing carbonyls.

## Conclusions

The three analytical procedures described in this chapter for measuring oil quality and/or oxidative stability may be useful under some conditions of testing. None of the methods is as sensitive as sensory evaluations done by a well-trained panel, but they are capable of verifying what is determined by the panel provided the analytical procedures are used and interpreted correctly. Because oxidation of fats and oils is so complex, there are many factors affecting the final outcome of the oil flavor. Therefore, measuring only one chemical change is not likely to predict the flavor quality of an oil. The CD, CV, and AV procedures measure general rather than specific compounds, so their flavor prediction falls short of ideal. Other procedures for measuring more specific flavor compounds in oils are discussed in other chapters of this book.

### Acknowledgment

This is Journal Paper No. J-15021 of the Iowa Agriculture and Home Economics Experiment Station, Ames, Iowa, Project No. 2568.

### References

1. Farmer, E.H., Koch, H.P., and Sutton, D.A. 1943. *J. Chem. Soc.* 541.
2. Noor, N., and Augustin, M.A. 1984. *J. Sci. Food Agric. 35:*805.
3. *Official and Tentative Methods of the American Oil Chemists' Society, 4th edition,* edited by D. Firestone, American Oil Chemists' Society, Champaign, IL, 1989.
4. Privett, O.S., and Blank, M.L. 1962. *J. Am. Oil Chem. Soc. 39:*465.
5. St. Angelo, A.J., Ory, R.L., and Brown, L.E. 1975. *J. Am. Oil Chem. Soc. 52:*34.
6. Tautorus, C.L., and McCurdy, A.R. 1990. *J. Am. Oil Chem. Soc. 67:*525.
7. White, P.J., and Miller, L.A. 1988. *J. Am. Oil Chem. Soc. 65:*1334.
8. Miller, L.A., and White, P.J. 1988. *J. Am. Oil Chem. Soc. 65:*1324.
9. White, P.J., and Armstrong, L.S. 1986. *J. Am. Oil Chem. Soc. 63:*5.
10. Duve, K.J., and White, P.J. 1991. *J. Am. Oil Chem. Soc. 68:*365.
11. Liu, H.-R., and White, P.J. 1992. *J. Am. Oil Chem. Soc. 69:*528.
12. Liu, H.-R., and White, P.J. 1992. *J. Am. Oil Chem. Soc. 69:*533.
13. Yoshida, H., Hirooka, N., and Kajimoto, G. 1990. *J. Food Sci. 55:*1412.
14. Fishwick, M.J., and Swoboda, P.A.T. 1977. *J. Sci. Food Agric. 28:*287.
15. Parr, L.J., and Swoboda, P.A.T. 1976. *J. Food Technol. 11:*1.
16. Farmer, E.H., and Sutton, D.A. 1943. *J. Chem. Soc.* 119.
17. Henning, G.J. In *Objective Methods for Food Evaluation,* National Academy of Sciences, Washington, D.C., 1976, pp. 155-170.
18. Henick, A.S., Benca, M.F., and Mitchell, J.H., Jr. 1954. *J. Am. Oil Chem. Soc. 31:*881.
19. *Standard Methods for Analysis of Fats and Oils,* Japanese Oil Chemists' Society, Nihonbashi, Chuo-ku, Japan, 1986.
20. Mizuno, G.R., and Chipault, J.R. 1965. *J. Am. Oil Chem. Soc. 42:*839.
21. Fioriti, J.A. 1965. *J. Am. Oil Chem. Soc. 42:*743.
22. Chu, Y.-H. 1991. *J. Am. Oil Chem. Soc. 68:*379.

23. Graziano, V.J. 1979. *Food Technol. 33:*50.
24. Fritsch, C.W., Egberg, D.C., and Magnuson, J.S. 1979. *J. Am. Oil Chem. Soc. 56:*746.
25. Paradis, A.J., and Nawar, W.W. 1981. *J. Food Sci. 46:*449.
26. Smith, L.M., Clifford, A.J., Hamblin, C.L., and Creveling, R.K. 1986. *J. Am. Oil Chem. Soc. 63:*1017.
27. Boki, K., Wada, T., and Ohno, S. 1991. *J. Am. Oil Chem. Soc. 68:*561.
28. Holm, U., Ekbom, K., and Wobe, G. 1957. *J. Am. Oil Chem. Soc. 34:*606.
29. Holm, U. *Abstracts,* International Society for Fat Research Congress, Yoteberg, Sweden, June 1972.
30. *Standard Methods for the Analysis of Oils, Fats and Derivatives,* Applied Chemical Division, Commission on Oils, Fats, and Derivatives, 7th edition, edited by C. Paqout, Pergamon Press, New York, 1987.
31. Johansson, G., and Persmark, V. 1971. *Oil Palm News 10/11:*3.
32. List, G.R., Evans, C.D., Kwolek, W.F., Warner, K., and Boundy, B.K. 1974. *J. Am. Oil Chem. Soc. 51:*17.
33. Snyder, J.M., Mounts, T.L., and Holloway, R.K. 1991. *J. Am. Oil Chem. Soc. 68:*285.
34. Schnepf, M., Spencer, G., and Carlat, J. 1991. *J. Am. Oil Chem. Soc. 68:*281.
35. Snyder, J.M., and Mounts, T.L. 1990. *J. Am. Oil Chem. Soc. 67:*800.
36. Idris, N.A., Abdullah, A., and Halini, A.H. 1992. *J. Am. Oil Chem. Soc. 69:*272.

# Chapter 10

# Accelerated Stability Methods

**Peter J. Wan**

Southern Regional Research Center, ARS, USDA, 1100 Robert E. Lee Boulevard, New Orleans, Louisiana 70124-4305, USA.

## Introduction

Edible oils and fats are mixtures of triglycerides and numerous minor constituents, such as sterols, tocopherols, free fatty acids, mono- and diglycerides, natural pigments, phosphatides, and trace metals. The chemical composition and physical properties of an oil can be influenced by the growing conditions and agricultural practices to which the oil source has been subjected. In addition, oils have diverse applications in foods and food preparation as well as industrial products. Therefore, it is not feasible to describe the quality of these materials by a single chemical or physical factor. In practice, a selected group of properties is used to define and assess the quality of a fat or oil. The limits for each of the selected properties are defined by the users and producers of fats and oils, and they are used as the specifications of the oil product for trade. Table 10.1 lists the sample specifications of several oils and shortenings (1,2).

During the shipping, handling, and storage of an oil, one or more of the properties specified can change by various degrees and rates. Hence, the keeping quality or shelf life of an oil is defined as the length of time that the oil or fat resists significant change of its acceptable characteristics. Theoretically, the shelf life or stability of an oil product should be determined by periodic examination of selected properties to ensure that the specific properties are within acceptable limits during storage and use. Occasionally, a performance evaluation of the oil product in a baking or frying application is conducted to assess the quality changes. This effort is time consuming and labor intensive. A well-packaged salad oil usually will need months of storage at ambient temperatures before any sign of deterioration, such as oxidative rancidity, peroxide value increase, or discoloration, develops. A shortening held under similar storage conditions will take much longer to develop significant degradation. Even a small-scale frying performance test can take several weeks to complete. Therefore, other than an occasional full-scale performance evaluation for fat quality and stability, regular quality assurance and quality control (QA/QC) of oil products often relies on accelerated tests which have been developed since the 1930s to quickly assess the stability of fats and oils.

Accelerated stability tests are selected based on the end use of each oil product. Since oxidation is the primary cause of oil degradation (3,4), most of the accelerated tests are designed to expedite the oxidation process by exposing oil samples to

**TABLE 10.1**
Specifications of Oils and Shortenings

| | Oil | | | | | |
|---|---|---|---|---|---|---|
| Typical Analysis | Corn | Soybean | Cottonseed | Peanut | Frying Liquid | Donut Shortening |
| Melting point (°C) | — | — | — | — | 41.1 | 46.1 |
| AOM stability (hr) | 10 | 8 | 10 | 15 | 35 | 60 |
| Smoke point (°C min.) | 232 | 232 | 232 | 227 | 227 | 227 |
| Cold test (hr) | Does not cloud | Does not cloud | 20 | — | N/A | N/A |
| Free fatty acid (% max.) | 0.05 | 0.05 | 0.05 | 0.05 | 0.05 | 0.05 |
| Peroxide value (me/Kg, max.) | 0.5 | 0.5 | 0.5 | 0.5 | 0.5 | 1.0 |
| Iodine value | 125 | 130 | 112 | 90-110 | 102 | 76-79 |
| Lovibond color (red, max) | 4.0 | 2.0 | 2.0 | 2.0 | 2.0 | 2.0 |
| Odor and flavor | Bland | Bland | Bland | Slightly nutty | Bland | Bland |
| Solid fat index (%) | | | | | | |
| 10°C | | | | | 4-6 | 30 |
| 21.1 | | | | | 3-4 | 21 |
| 26.7 | | | | | — | 18 |
| 33.3 | | | | | 2-3 | 13 |
| 40 | | | | | 1-2 | 9 |

elevated temperatures in the presence of excess amounts of air or oxygen. In some cases, intense light or subambient temperature is applied to promote the chemical reactions or physical changes. The progress of oil degradation can be monitored by measuring either the disappearance of oxygen or the amount of oxidized products generated. The course of oil oxidation usually proceeds slowly at the initial stage and almost always is followed by a sudden rise in oxidation rate after a certain period of time which is called the induction period (IP) or induction time (IT). This phenomenon can be described graphically by the curve shown in Figure 10.1. In this chapter, both classical and recently developed accelerated methods are reviewed.

## Schaal Oven Test

This is a simple accelerated stability test of oil and fatty foods for their keeping quality (5,6). Normally, 50 g of oil or a fatty food sample is stored in a loosely sealed glass container and placed in an oven maintained at 63°C. The quality of the oil sample is examined periodically by a sensory panel until a definite rancid odor or flavor develops. The number of days required to develop rancidity is recorded as the end point of the sample tested. The procedure can be modified to instruct panelists testing for any off-flavor development.

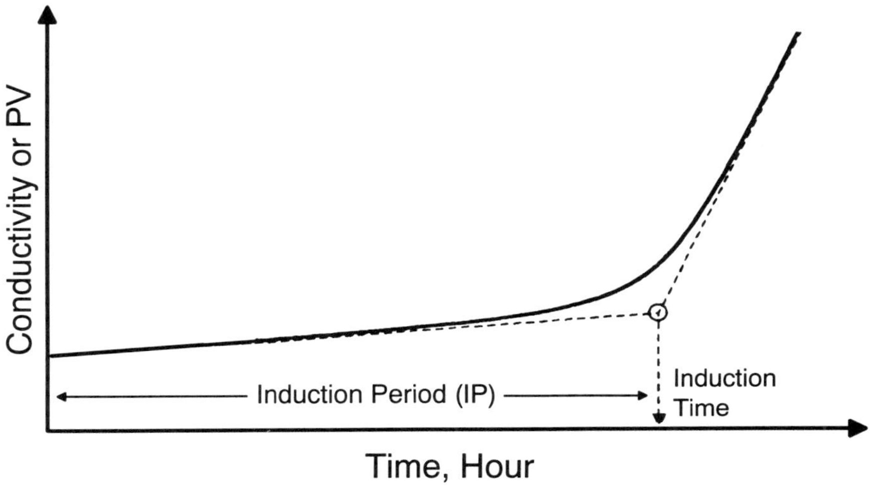


**Figure 10.1.** Course of oil oxidation in conductivity or peroxide value (PV) versus time.

Since no official standard aside from sensory evaluation is assigned to this common method, many variations in testing conditions (5) and different objective assays (7–9) were adopted to determine the end point for the Schaal Oven Test. For example weight gain or peroxide value (PV [7,8]); anisidine value (AV); Totox value, equal to 2 × PV + AV (9); and total volatiles by gas chromatography (10) have been used. Each of these methods has its inherent sensitivity and reproducibility for end point determination. Reproducibility of ±0.4 d or 11.1% was reported for 528 samples of sunflower, zero erucic rapeseed (canola), and soybean oils when either PV or gravimetrical measurement was used as the end point determination (8).

### *Advantages and Disadvantages*

The Schaal Oven Test is a simple approach to the oil stability problem. It requires minimal laboratory equipment when sensory evaluation is used to define the end point. However, the test requires a large sample and a relatively long time (4-8 d) to complete. Therefore, even though the data correlates well with shelf stability of oil, use of this method as a routine test is rare unless sensory panels are available.

## Active Oxygen Method (AOM)

Active oxygen method (AOM) was designed to measure the peroxides formed by reaction of active oxygen with oil (11–16). The AOM, also known as the Swift stability test (13,14,17) is the most popular accelerated stability test for oils and is an official AOCS Method (Cd 12-57 [18]). The 20 mL oil sample is placed in a glass tube and is aerated continuously at 97.8°C. One gram samples are withdrawn periodically for PV determination. The number of hours required for the oil sample to reach a peroxide value of 100 mEq/kg (milliequivalents of peroxides per kg of oil) is reported as the AOM hours for the oil.

### *Advantages and Disadvantages*

Because this method requires periodic titration to determine PV of an oil, it is a time-consuming procedure. Minimizing the frequency of sampling reduces both accuracy and reproducibility of the method. The official AOCS Method, Cd 12-57, demands a minimum of three peroxide titrations per sample. Results may have a standard deviation of ±13.4% or a maximum variation between laboratories of ±25 hr on a 100-hr sample or 2.5 hr on a 10-hr sample. Active Oxygen Method is often used since it is an established AOCS official method and an important specification of fats and oil products traded in the marketplace (Table 10.1). Large variations in AOM hours for an oil sample can be minimized by frequent titration.

## Oil Stability Index (OSI)

Oxidative stability of oils and fats can be determined automatically by commercially available equipment, Rancimat from Brinkmann Instrument (Westbury, NY [19]) or Oxidative Stability Instrument from Omnion, Inc. (Rockland, MA [20]). Instead of measuring peroxides as in AOM, this method monitors the volatile compounds generated from an oil sample heated under constant aeration. These volatile compounds are primarily short-chain organic acids, such as formic and acetic acids, which develop in thermally oxidized oil (21,22). Volatiles from heated oil are trapped in water, and their rate of formation is monitored by electro-conductivity. The end point or induction time determined is the oil stability index (OSI) for the specific oil at the selected temperature. This method correlated well with the AOM test with coefficients of 0.987 and 0.976 at 100 and 110°C, respectively (23).

The following experiment uses a Metrohm Rancimat (Model 679) and requires as little as 2.5 g oil, but normally 5 g is used. The oil sample is placed in a reaction cell and kept at a constant temperature with an air flow of 20 L/hr through the oil. The purged air stream is bubbled into distilled water (60 mL) in a bottle equipped with a conductivity electrode. The conductivity data of the heated oil sample is plotted against time as shown in Figure 10.1. This data is stored electronically, and conductivity is checked periodically. A baseline slope is generated when a rapid change in conductivity is detected. After the conductivity rise plateaus, a second slope is established. The intercept of these two slopes, derived automatically by the Rancimat software, is reported as the induction period or OSI in hours for the oil at the selected temperature. A Rancimat can run six samples simultaneously at a single temperature between 50 and 180°C.

The OSI instrument manufactured by Omnion, Inc. uses a similar design principle as that of a Rancimat, and its induction time is determined by using second derivatives of the conductivity versus time data. The automated OSI test, using instruments like the Rancimat or Omnion Instrument, has been recently adopted by the American Oil Chemists' Society as an official method (Cd 12-b92 [18]) to replace AOM.

### *Advantages and Disadvantages*

A very good correlation with a slope of 1.005 and correlation coefficient of 0.987 between OSI data and AOM was obtained from a Rancimat for peanut, sunflower, and olive oils; lard; margarine; and cooking butter (23). Oil Stability Index data for several vegetable oils collected using a Rancimat had good reproducibility and linearity between Log(OSI) and temperature (24). This automated method offers an efficient alternative to the conventional AOM test and may be used to assess the effects of refining processes, impurities, and antioxidants on the keeping quality of oils. It may also assist in the selection of optimal oil blends and formulations for specific uses.

A disadvantage of OSI is that fat-containing foods are difficult to test directly. Therefore, isolation of the fat from these samples is necessary (25). However, work by Barrera-Arellano and Esteves (26) demonstrated the feasibility of using the Rancimat to determine oxidative stability of potato chips without oil extraction. Due to the presence of volatile compounds in fried potato chips, they observed two inflection points in their Rancimat results. Their sensory results showed a correlation coefficient of 0.86 with the induction time of the second inflection point. A disadvantage of high-temperature accelerated tests pointed out by Frankel (27) was that oxidation of oils at elevated temperatures (100°C and above) sometimes differs from that at lower temperatures. Therefore, it is always advisable to compare the relative stability of various oils at different temperatures.

## Oxygen Bomb Method (OBM) and Other Oxygen Consumption Methods

For stable fats, such as frying shortenings, more than 2 days are needed for an AOM test. The oxygen bomb method (OBM) is designed to further accelerate the stability test for relatively stable fats (28,29). In this method, a 10 g shortening sample is held at 135°C in a glass-lined reactor under an initial oxygen pressure of 110 psig. The experiment is carried out until a sudden drop of oxygen pressure, usually 2 psig, is recorded. Oxygen Bomb Method is reported in minutes and often converted to AOM value based on an empirical correlation between AOM and OBM:

$$\text{AOM (hr)} = \text{Slope} \times \text{OBM (min)} + \text{Intercept}$$

Other oxygen uptake tests were also reported to have excellent correlations with AOM (30) and OSI (31). A new instrument designed according to the principle of FIRA-Astell Apparatus (32) was made available by Mikrob Aarhus from

Denmark and is called an Oxidograph. This unit can simultaneously run six samples. Each reaction vial uses 5-g samples and 100 mL oxygen. The oxygen consumption is monitored continuously at 100°C. Its end point determination is similar to that of OBM and essentially an inverse graphic of OSI.

### *Advantages and Disadvantages*

Although these oxygen consumption methods are not as popular as the AOM method, they offer the following advantages: OBM is 2–10 times faster than the AOM test (33); oxygen pressure is monitored automatically, providing better reproducibility than AOM; and fat-containing foods such as potato chips, and raw and roasted peanuts (33), or oils containing volatile antioxidants, such as BHA and BHT, can be tested directly by these oxygen consumption methods. The OBM was also shown to correlate well with shelf life and Schaal Oven test of oils (34). Maintaining the high oxygen pressure and temperature for OBM may require some special attention. The Oxidograph with its current design may not be suitable for samples with induction periods longer than 24 hr or samples containing moisture at temperatures approaching 100°C.

## Light Exposure Test

The deleterious effect of light on the flavor stability of edible oils (35–43) and various fat-containing foods, such as milk (44–47), potato chips (48), and walnut meats and oil (49), is well characterized. Light-induced or light-struck flavor from an oil is often described as light reversion flavor, which has a characteristic grassy flavor and astringent mouthfeel. Infrared light can produce a grassy reversion flavor rapidly, while ultraviolet will induce an astringent mouthfeel, such as a drying sensation (36,37). Dark amber glass bottles effectively retard off-flavor development in edible oils exposed to diffused light but not to direct sunlight (39,40). Although there is no official method for light exposure testing, a reproducible light exposure test was suggested by Moser et al. (41) and has been demonstrated by Warner et al. (42,43).

The experimental conditions used by Moser (41) and Warner (42) were to place the oil in an 8-oz, narrow-mouthed, clear glass bottle with air in the headspace and loosely stopper with a cellophane- or teflon™-covered cork, then expose the samples to fluorescent light at 7535 lux (700 ft candles) at 30°C. Oil quality is evaluated by a trained sensory panel every 2 or 4 hr until off-flavors characteristic of light exposure are detected. This method has a reproducibility of ±0.8 for a flavor score from 1 (bad) to 10 (excellent [41]). Gas chromatographic volatiles can also be used to evaluate the light stability of oils (43).

### *Advantages and Disadvantages*

This is an effective accelerated stability test to assess the deleterious effect of light on the quality of oils. It can be used to assist in the selection of packaging material for oil (42). But detection of the typical light-struck flavor of oil requires a trained sensory panel. The amount of oil required plus the cost of training and maintaining a sensory panel have restricted the use of this procedure.

## Smoke Point

The smoke point is the temperature at which an oil, heated at a rate of 5–6°C/min, will begin to emit continuous wisps of smoke. Procedural details can be found in AOCS Method Cc 9a-48 (18). The free fatty acid content of an oil inversely correlates with the smoke point, as shown in Table 2 (50). Therefore, this method serves as a quick check of residual free fatty acids or other nontriglyceride components, such as monoglycerides, in a finished fat and oil product. A good frying shortening should have less than 0.4% monoglycerides and less than 0.05% free fatty acid, and will have a smoke point of 218°C or higher.

### *Advantages and Disadvantages*

This is a simple, qualitative method, but it requires some practice to achieve the desired reproducibility and reliability.

## Cold Test

The cold test, official AOCS Method Cc 11-53 (18), is an accelerated test to measure the resistance of an oil sample to crystallization under subambient temperatures. The cold test data is used as an index of the winterization and stearin removal processes, and gives an indication of how long an oil will keep in a refrigerator before cloudy fat crystals form. In this test, a prefiltered 200-300 mL oil sample is heated to 130°C. A portion of the oil is poured into a 115-mL (4-oz) clear glass bottle fitted with a cork stopper. After adjusting the sample temperature to 25°C and sealing the cork with paraffin, the entire bottle is immersed in an ice water bath. After 5.5 hr and every hour thereafter, the oil sample is examined for fat crystals or

**TABLE 10.2**
Effect of Free Fatty Acid on Smoke Point of Shortenings

| Free Fatty Acid, % | Smoke Point,°C |
|---|---|
| 0.04 | 218 |
| 0.06 | 210 |
| 0.08 | 205 |
| 0.10 | 200 |
| 0.20 | 190 |
| 0.40 | 177 |
| 0.60 | 171 |
| 0.80 | 165 |
| 1.00 | 160 |

cloudiness. The test ends when the cloudiness is first detected and results are reported in number of hours. Typical data are shown in Table 10.1.

### *Advantages and Disadvantages*

The cold test is a simple accelerated test routinely applied in the trade, but it requires practice to achieve acceptable reproducibility. Presently, there is no automated testing procedure available for this method.

## Thermal Analysis

### *Oxidative Stability of Oil*

Application of thermal analysis as accelerated oil stability tests has been studied by many researchers (51–57). Cross (51) used a standard Differential Scanning Calorimeter (DSC) at 150°C with a purged oxygen rate of 20 mL/min to measure the stability of oils. The end point of DSC was taken at the time where a rapid exothermic reaction of oil and oxygen occurred. Testing of a 14-d AOM oil can be replicated by DSC in less than 4 hr. The correlation coefficient between DSC data and AOM was 0.974. Hassel (53) demonstrated that Pressure Differential Scanning Calorimetry (PDC) can further accelerate the oil stability test of a standard DSC. At 120°C and 500 psig oxygen pressure, the stability test can be done in minutes instead of hours, with a relative standard deviation of 10%. The classic rancidity test, for example monitoring the weight gain of oil in a Schaal oven test, can be accomplished easily and efficiently with Thermogravimetric Analysis (TGA) procedures (53). Thermogravimetric analysis as an accelerated oil stability test can be carried out isothermally or in a temperature-programmed mode (53,54). Under isothermal conditions (137°C, 50 mL $O_2$/min), the induction time or end point of an oil sample is determined by the rapid weight gain. In a temperature-programmed or dynamic mode (5°C/min, from 65°C, 50 mL $O_2$/min), the onset temperature where rapid weight gain begins is the end point of the test. Either mode has a relative standard deviation <10% and a good correlation with AOM. In addition to the work by Hassel (53), TGA as an oil stability test has been examined separately by other scientists (55-57).

### *Crystallinity of Shortening*

Differential Scanning Calorimetry is one of the most versatile tests used to study the quantity and quality of fat crystals in shortening and margarine. From the melting thermogram of a sample, one can estimate the types and proportions of fat crystals present, namely, α, β', and β. β' crystals are associated with smooth mouthfeel and even spreading, and therefore are considered the most desirable polymorphic for margarine and shortenings. The solid fat content (SFC) of a sample at a temperature within its melting range can be estimated from its DSC melting thermogram.

The solid fat content determined by DSC is comparable to the solid fat index (SFI) measured with dilatometry (58,59). Fat crystals are known undergo changes during processing, tempering, or storage (60,61). The desired β' crystal may be formed continuously and later transformed to a more stable β form. This dynamic nature of fat crystal stability of a shortening can be assessed with DSC under conditions similar to normal processing, tempering, or storage temperature profiles (61).

### *Advantages and Disadvantages*

The requirements of sophisticated instrumentation and trained operators may be the major obstacles to using thermal analysis for QA/QC of fats and oils. The advantages of using thermal analysis are multiple. It requires a very small sample, usually 10–400 mg. Besides stability index, thermal analysis can provide several useful thermal properties of an oil, such as specific heat, melting point and range, and melting enthalpy. The melting enthalpy, the heat required to melt the sample, can be used to estimate SFC and phase-transition characteristics of fat crystals. Thermal analysis is sensitive and specific.

## Chemiluminescence (CL)

Many reactions proceed with the evolution of heat, but some emit light. Measurement of this ultraweak chemiluminescence (CL) has long been the focus of scientists. In the late 1960s and early 1970s, Battelle Memorial Institute conducted experiments to develop the CL technique and instrument (62) and to apply it to monitor the oxidation of fats and oils. Usuki et al. (63) followed vegetable oil oxidation at elevated temperatures by CL in qualitative comparison with conventional methods.

More recently, Cash et al. (64) used CL to study the oxidation of sunflower oil. They used 40 μL oil samples in small aluminum thermal analysis pans placed in the center of the CL sample chamber. The photon counter was set to read the CL emission every 100 sec while the sample was kept at a constant temperature between 90 and 121°C. The plot of CL emission (Hz) versus time resembles the characteristics of an OSI plot. The induction time was determined as the point of rapid increase in CL emission. An Arrhenius activation energy plot of ln(Induction Time) vs. reciprocal absolute temperature provided a predicted induction period at 25°C of 1.21 yr and an activation energy of 93 kJ/mole for sunflower oil (64).

### *Advantages and Disadvantages*

Chemiluminescence is a very sensitive method. Its photon counter can detect reaction rates as low as $10^{-9}$ mole/yr for solid samples and $10^{-14}$ mole/yr for liquid samples (62). But very little CL data on oil stability has been reported. Due to the increasing availability of commercial luminometer systems (65), we should expect renewed interest in its application to the oil quality and stability study.

## References


1. Swern D. *Bailey's Industrial Oil and Fat Products,* 4th Ed., John Wiley & Sons, New York, 1982.
2. Weiss, T.J. *Food Oils and Their Uses,* Avi Publishing Company, Inc., Westport, CT, 1983.
3. Gunstone, F.D. 1984. *J. Am. Oil Chem. Soc. 61:*441.
4. Gray, J.I. 1978. *J. Am. Oil Chem. Soc. 55:*539.
5. Joyner, N.T., and McIntyre, J.E. 1939. *Oil and Soap 15:*184.
6. Cocks, L.V., and van Rede, C. *Laboratory Hand Book for Oil and Fat Analysis,* Academic Press, New York, 1966, pp. 304-41.
7. Olcott, H.S., and Einset, E. 1958. *J. Am. Oil Chem. Soc. 35:*161.
8. Pokorny, J., Dobiasova, S., and Dvidek, J. *Sbornik; Vysoke Skoly Chemicko. Tech. v Praze, E;* No. 58, p. 163, 1985.
9. Castro, M.E., Masson, S.L., and Valenzuela, G.F. 1986. *Alimentos 11:*5.
10. Warner, K., and Frankel, E.N. 1985. *J. Am. Oil Chem. Soc. 62:*100.
11. Paschke, R.F., and Wheeler, D.H. 1944. *Oil and Soap 21:*52.
12. Wheeler, D.H. 1932. *Oil and Soap 9:*89.
13. Kilgore, L.B. 1933. *Oil and Soap 10:*66.
14. King, A.E., Roschen, H.L., and Irwin, W.H. 1933. *Oil and Soap 10:*105.
15. King, A.E., Roschen, H.L., and Irwin, W.H. 1933. *Oil and Soap 10:*204.
16. Riemenschneider, R.W., Turer, J., and Speck, R.M. 1943. *Oil and Soap 20:*169.
17. Fore, S.P., Moore, R.N., and Bickford, W.G. 1951. *J. Am. Oil Chem. Soc. 28:*73.
18. *Official Methods and Recommended Practices of the American Oil Chemists' Society,* 5th Ed., Champaign, IL 1992.
19. Frank, J., Geil, J.V., and Freaso, R. 1982. *Food Tech. 36:*71.
20. Jebe, T.A., Matlock, M.G., and Sleeter, R.T. 1993. *J. Am. Oil Chem. Soc. 70:*1055.
21. Pardun, H., and Kroll, E. 1972. *Fette-Seifen-Anstrichmittel 74:*366.
22. deMan, J.M., Fan, T., and deMan, L. 1987. *J. Am. Chem. Soc. 64:*993.
23. Laubli, M., and Bruttel, P.A. 1986. *J. Am. Chem. Soc. 63:*792.
24. Hassenhuettl, G.L., and Wan, P.J. 1992. *J. Am. Chem. Soc. 69:*525.
25. Sandmeier, D., and Ziegleder, G. 1985. *Suesswaren 29:*487.
26. Barrera-Arellano, D., and Esteves, W. 1990. *J. Am. Oil Chem. Soc. 69:*335.
27. Frankel, E.N. 1993. *Trends Food Sci. Tech. 4:*220.
28. Beasley, R.W., and Sampson, R.L. *Modification of the Oxygen Bomb Stability Technique,* presented at Cereal Chemists' Meeting, May 4, 1960.
29. Bennett, J.E., and Byer, M.J. 1964. *J. Am. Oil Chem. Soc. 41:*505.
30. Nagao, A., and Yamazaki, M. 1981. *J. Japan Oil Chem. Soc. 30:*778.
31. Larsen, K.E. *Proceedings of 15th Scandinavian Symposium on Lipids,* Rebild Bakker, Denmark, June 11–15, 1989.
32. Rossell, J.B. 1989. *Food Sci. Tech. 3:*235.
33. Blankenship, B.R., Holaday, C.E., Barnes, P.C., Jr., and Pearson, J.L. 1973. *J. Am. Chem. Soc. 50:*377.
34. Pohle, W.D., Gregory, R.L., Weiss, T.J., Van Giessen, B., Taylor, J.R., and Ahern, J.J. 1964. *J. Am. Oil Chem. Soc. 41:*795.
35. Coe, M.R., and LeClere, J.A. 1934. *Ind. Eng. Chem. 26:*245.
36. Golumbic, C., Martin, C.J., and Daubert, B.F. 1946. *Oil Soap 23:*187.

37. Golumbic, C., Martin, C.J., and Daubert, B.F. 1946. *Oil Soap 23:*360.
38. Gudheim, A.R. 1943. *Oil Soap 20:*197.
39. McConnell, J.E.W., and Esselen, W.B., Jr. 1946. *Oil Soap 23:*369.
40. McConnell, J.E.W., and Esselen, W.B., Jr. 1947. *J. Am. Oil Chem. Soc. 24:*6.
41. Moser, H.A., Evans, C.D., Cowan, J.C., and Kwolek, W.F. 1965. *J. Am. Oil Chem. Soc. 42:*30.
42. Warner, K., and Mounts, T.I. 1984. *J. Am. Oil Chem. Soc. 61:*548.
43. Warner, K., and Frankel, E.N. 1987. *J. Am. Oil Chem. Soc. 64:*213.
44. Anderson, K.P. 1959. *Intern. Dairy Congr.,* 15th (London) *3:*1746.
45. Dunkley, W.L., Franklin, J.D., and Panghorn, R.M. 1962. *Food Tech. 16:*112.
46. Samuelsson, E.G., and Thorne, K.E. 1959. *Intern. Dairy Congr.,* 15th (London) *1:*488.
47. Fellman, R.L., Dimick, P.S., and Hollender, R. 1991. *J. Food Protect. 54:*113.
48. U.S. Department of Agriculture. "Manufacture of Potato Chips," *Bur. Agr. Ind. Chem.,* Washington D.C., 1945.
49. Musco, D.D., and Cruess, W.V. 1954. *J. Agr. Food Chem. 2:*520.
50. Swern, D. In *Bailey's Industrial Oil and Fat Products,* 3rd Ed, Interscience Publishers Div., John Wiley & Sons, New York, p. 122, 1964.
51. Cross, C.K. 1970. *J. Am. Oil Chem. Soc. 47:*229.
52. Blaine, R.L. 1974. *Amer. Lab. 6:*18.
53. Hassel, R.L. 1976. *J. Am. Oil Chem. Soc. 53:*179.
54. Nieschlag, H.J., Hagemann, J.W., and Rothfus, J.A. 1974. *Anal. Chem. 46:*2215.
55. Buzas, I., and Kurucz, E. 1979. *J. Am. Oil Chem. Soc. 56:*685.
56. Mikula, M., and Khayat, A. 1985. *J. Am. Oil Chem. Soc. 62:*1694.
57. Takaoka, K., and Kobayashi, K. 1986. *J. Am. Oil Chem. Soc. 63:*1447.
58. Suzuki, K., Murase, I., Nagai, T., Marugama, T., Niiya, I., Imamura, M., and Matsumoto, T. 1973. *J. Japan Oil Chem. Soc. 22:*732.
59. Lambelet, P., Desarzens, C., and Raemy, A. 1986. *Lebensmittel-Wissenschaft und Tech. 19:*77.
60. Blaurock, A.E., and Carothers, A.P.R. 1990. *Inform 1:*351.
61. Blaurock, A.E., and Wan, P. 1990. *Inform 1:*351.
62. Nathan, R.A., Mendenhall, G.D., Hassell, J.A., and Wallace, J.D. 1975. *Indust. Res. 17:*62.
63. Usuki R., Kaneda, T., Yamagishi, A., Takyu, C., and Inaba, H. 1979. *J. Food Sci. 44:*1573.
64. Cash, G.A., George, G.A., and Bartley, J.P. 1988. *J. Sci. Food Agric. 43:*277.
65. Stevenson, R. 1992. *Am. Biotech. Lab.* Oct. p. 11.

Chapter 11

# Applied Statistics for Oil Chemists

**Terry C. Nelsen**

USDA-Agricultural Research Service-Midwest Area, 1815 North University St., Peoria, IL 61604, USA.

## Introduction

The purpose of this chapter is to make the reader a more intelligent consumer of statistics. A scientist or technician is not expected to be an accomplished statistician, but he or she should have an understanding of some basic statistical methods, what the underlying assumptions associated with these methods are, and what the results mean in a biological or chemical sense. Statistics was developed to enable us to make general statements about a population of objects or items and to measure the effects of specific influences (treatments) on populations or parts of populations. Engineers have given us the concepts of signals and noise, and much of what statistics does is to identify and measure signals in the presence of noise. Most of the time, the signals are the effects of our treatments and the noise is variation from all other sources.

This chapter is not meant to be a minicourse in statistics. Many good books are available for basic statistical theory and specific applications. *Statistical Methods* by Snedecor and Cochran and *Principles and Procedures of Statistics* by Steel and Torrie are widely used in the Agricultural Sciences. *Statistics in Research* by Ostle and *Probability and Statistics for Engineers and Scientists* by Walpole and Myers effectively explain elementary concepts and applications. Aspects of experimental design are covered by Hicks in *Fundamental Concepts in the Design of Experiments*; and Box, Hunter, and Hunter cover a great deal about factorial designs in *Statistics for Experimenters*. More specialized applications to chemistry are found in *Chemometrics: A Textbook*, edited by Massart et al. and to analytical chemistry in *Use of Statistics to Develop and Evaluate Analytical Methods* by Wernimont.

This chapter is designed to review basic concepts of statistics and to present these concepts in a thought-provoking manner. When I review papers I often come across statements such as "the results from treatment A were different from the results from treatment B, but the difference was not statistically significant." My first thought is "why wasn't the difference significant?" Did the researcher use the correct test? Did the variation overwhelm the treatment effect? Was an unknown factor influencing the data? Were the number of observations insufficient? The previous statement should be immediately followed by an explanation of why the difference was not "significant."

Many scientists are uneasy about statistics. Shouldn't we be able to find an answer rather than estimate it? Water has a specific boiling point, so why shouldn't soybeans have a specific oil content? The answer is obviously that while water can generally be thought of as a constant, a soybean is a product of the genetics of its parents and the many variables in the environment in which it was grown. As long as we deal with living organisms, we are going to have to deal with differences among them. Statistics is one way that we have of measuring, adjusting, and understanding variation in animals, plants, and their products.

In this chapter I will discuss some basic statistics and statistical concepts minimizing formula and Greek symbol use. I will present examples, but I do not expect this chapter to be used as a cookbook reference for specific tests or designs. I hope to cause the reader to think about why specific tests are used, what assumptions are made, what the consequences of violating these assumptions are, and what the probability statement means when it is presented with a statistic. I will discuss numbers in general, distributions with means and standard deviations (SD), probability, hypothesis testing, correlation, regression, analysis of variance, multivariate analysis, and finish with a general discussion of planning experiments. I don't plan to define "heteroscedasticity" or "multicolinearity" but I consider them to be fine words, and after you have read this chapter you may feel free to use them in casual conversation so that people will think you are a genuine statistician.

## Numbers

We use numbers to represent simple and complex things. We seldom think about exactly what these numbers represent. Many numbers are simple in measurement and in concept. The difference between 2 g and 3 g is 1 g and the difference between 9 g and 10 g is 1 g. Ten centimeters is twice as long as 5 cm. The average of 2 L, 4 L, 5 L, and 7 L is 4.5 L. Ratios can be a bit more complex. For example, the difference between one-half gram and one-third gram is one-sixth gram while the difference between one-ninth gram and one-tenth gram is one-ninetieth gram. Thus a one unit difference is not always simple when that one unit is in the denominator of a fraction. Is 10° twice as warm as 5°? Is the difference in temperature between 2° and 3° the same as the difference between 9° and 10°? (By the way, an increase from 5 to 10% is not a 5% increase, but a 100% increase. It is a 5 percentage point increase.) The average of 2 cats, 4 cats, 5 cats, and 7 cats is 4.5 cats. What is one-half of a cat? I hope it is a concept and not a measurement. Is a difference in pH of 1 the same if it is from pH 2 to 3 or from pH 7 to 8? Both a fraction and a logarithm are involved in measuring pH. Will a treatment that causes a "significant" change in pH in an acidic solution have the same effect in a basic solution?

Some of the most troublesome numbers are those from subjective measurements. Flavor scores and stage of growth or disease are common measurements that can be difficult to quantify and to calibrate. Any person assigning a measurement should be calibrated, just like any other instrument, if the measurements are to be reliable over time or between instruments. See Chapters 3 and 4 for valuable information on training and calibrating sensory judges.

These differences can be important when measuring or predicting the effects of a treatment. A measurement may be normally distributed but its reciprocal may not be. If a treatment will increase a measurement from 1 to 2%, will that same amount of treatment still cause a 9% measurement to increase to 10%? If we work with ratios, then we might expect a change of one-ninth to one-tenth to be harder to detect than a change of one-half to one-third. If x% of a volatile compound can cause a change in flavor score from 2 to 3, will two times x% cause a change to 4? Subjective measurements should be calibrated frequently to prevent order effects. (Does a score influence the next score? If I score from 1 to 10, will a 5 still be a 5 if a 1 or a 10 precedes it?) The calibration should be standard between instruments (i.e., people), labs, and years. Think about what the numbers represent and especially what changes in the numbers mean biologically or chemically.

## Distributions

Suppose we have measured the protein in 10 beans as 37, 39, 40, 40, 40, 41, 42, 43, 43, and 45%. We can say that the average amount of protein in these soybeans is "around 40%" which is often an adequate description. We may say that these soybeans have a protein content of "about 35–45%" which again can be an adequate description. Often, we require more rigorous descriptions. We can say that these beans had an average protein content of 41.0% and ranged from 37 to 45%. Here we used a central tendency (the average) and a measure of distribution (the range). As we get more specific in our description, we need to make sure that we are all using the same language.

The measures of central tendency are mean, median, and mode. The mean is simply the average of all the values in question. For our 10 beans the mean is 41.0%. The median is the value which has one-half of the data below it and one-half of the data above it. Arrange your values in order from smallest to largest. If you have an odd number of observations, the median is the middle value. If you have an even number of values, the median is the average of the two middle values. For our 10 beans the median is (40 + 41)/2 = 40.5%. A mode is a frequent value. Our beans have a mode at 40% and a smaller mode at 43%.

We are most familiar with the mean, but the median and mode can be helpful, especially in initial data analysis. The mean has the advantage of being widely understood, easy to calculate, and can be manipulated algebraically. The median is easy to calculate and usually is not influenced as heavily by extreme values as the mean. Note in our bean example if we exchanged the bean with 45% protein for a bean with 55% protein that the mean would change but the median would not. The median, however, is not easily manipulated algebraically, requires ordering the data, and is not as well understood by scientists. The median is most commonly used in skewed distributions where the average can be misleading. A common use is to describe salaries which have a few very high values (oil chemists) and many lower values (statisticians and other commoners). The mode (or relative mode) also requires ordering the data and is not as well understood as the mean. A population can have zero, one, two, or more modes. In quantitative data, a mode is a clustering of data around a point and does not have to be exact repetition of a value. The mode is most useful for scientists in initial data analysis and in some specialized cases. In the normal bell-shaped distribution the mean = the median = the mode. When you first look at your data, if the mean, median, and mode are not similar, then you may not have a normal distribution. If the mean and the median are quite different, the data may be skewed. If too much data is either above or below the mean, then you often can use a log form in order to bring your distribution into a normal configu-

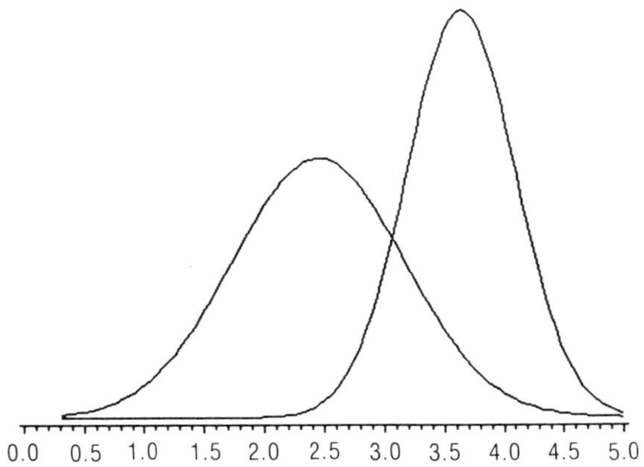


**Figure 11.1.** Normal distributions.

ration. Percentage or proportional data is often skewed but can be made to look normal with square-root or arcsine transformations.

Both of the distributions in Figure 11.1 are normal distributions. The distribution on the left has a mean of 2.45 and SD of 0.71, while the distribution on the right has a mean of 3.64 and SD of 0.45.

To convert your original normal distribution to a standard normal distribution (Figure 11.2), subtract the mean from every value and divide the result by the SD. The new distribution has a mean of 0 and SD of 1. The tables in statistics books refer to this new distribution.

When the statistical results just don't look right, the cause is often nonnormality of the original data. When you use normal statistics on nonnormal data, the estimates of variation will be inflated, and differences that look important will not be

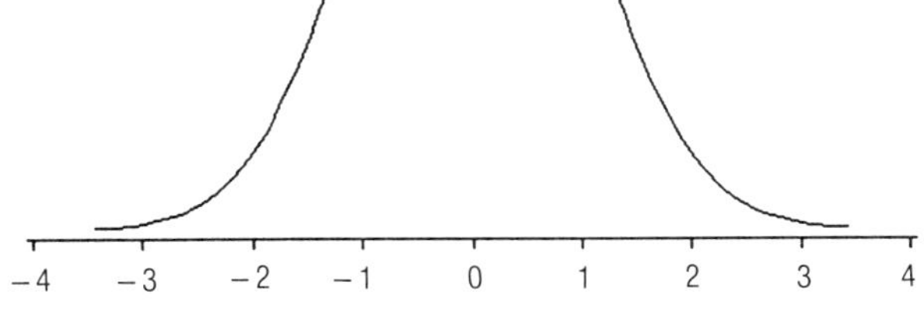


**Figure 11.2.** Standard normal distribution.

significant. The transformations mentioned above are commonly used to stabilize the variation. Transformation is not cheating, and a difference that is significant after transformation can be considered to be a real difference if you understand that you have changed the underlying distribution. Transformations should not be considered as anything worse than changing from inches to centimeters or from Fahrenheit to Centigrade. The consequences of not transforming and proceeding to analyze nonnormal data as if it were normal are an inflated estimate of variance and subsequent difficulties in recognizing and measuring real differences. If nonnormality is caused by a few very large or very small values then transformation is preferable to calling these oddball values "outliers" and simply dropping them from the analysis. Declaring values "outliers" and dropping them from analysis should be a last resort.

Distribution S1 in Figure 11.3 has a tail of data at the higher end. This tail is common in many microbial growth or contamination studies where some very large values can be measured. A log transformation is usually helpful. Distribution S2 where the upper end of the distribution is constrained, such as a 100% limit or a maximum flavor score, is also common.

If the data were randomly scattered along a range, the mean and the median could be similar, but modes may not be apparent or may be off-center. Another distribution can have mean = median = mode and still not be a normal distribution if an excess of data is grouped around the mean or at the extremes. This condition is called kurtosis and is often curable with transformations.

Distribution K1 in Figure 11.4 is common where only a very narrow range of the distribution has been chosen, or where an effort was made to choose a "representative" rather than a random sample of the data. Care was taken to include some data near the mean (but not too much) and some data near the extremes. The result is non normality and an overestimate of the true variance. Distribution K2 is common in biological data where the extremes don't survive or where the experimenter has taken pains to exclude any extremes. The result is an underestimate of the true variance.

A range tells you the total dispersion of the data but does not give any insight into how the data is distributed within this range. The most widely used measure of

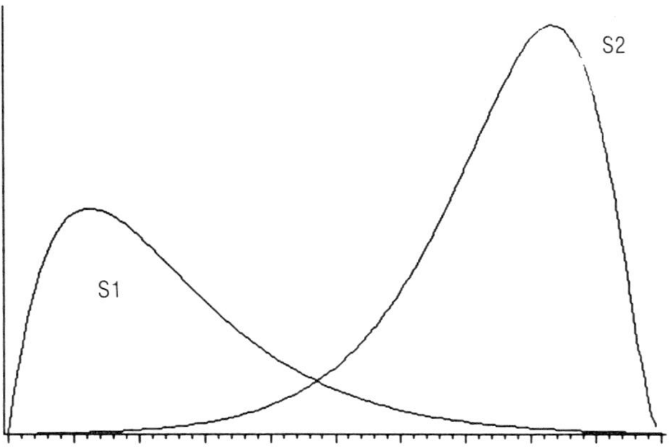


**Figure 11.3.** Skewed distributions.

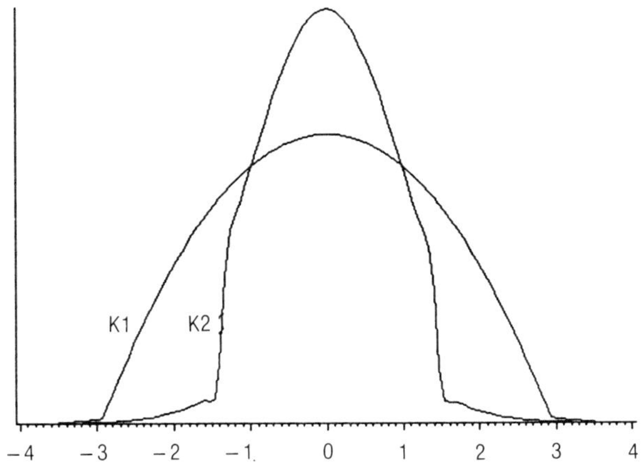


**Figure 11.4.** Kurtosis.

dispersion is the SD and its square, the variance. In a normal distribution, two-thirds of the data is within 1 SD of the mean, 95% is within 2 SD of the mean, and 99% is within 3 SD of the mean. A normal distribution can be used to describe a remarkable number of populations in nature. If you have a data set that is absolutely not normally distributed, you should consult either a statistician or a book on non parametric statistics. If you are concerned about the mean of a population rather than the population itself, you can make assumptions about the mean (and other means which you may be comparing it against) based on its normal distribution. The Central Limit Theorem states that if you take samples from a population and estimate a mean for each sample, these estimates will form a normal distribution. Thus, if you are comparing two populations, you must consider the distributions of each. If, however, you are comparing the means of the two populations, then you can use standard assumptions and tests based on the normal distribution of each.

### *Example 1: Mean and Variance*

The data in Table 11.1 is hypothetical data from a sample of 12 beans. Using our keen powers of observation and a few statistics, what can we decide about these beans?

The first thing to do in any data analysis is to look at the data. (In our modern computer age there is no excuse for not looking at your data. Very large data sets can be condensed or sampled.) For a small data set like Table 11.1, an effective visual is often a simple plot along a line (see Figure 11.5). There are no rigid rules for this type of data presentation; you are doing this for your own benefit, and you can make the scales and use whatever symbols are most helpful to you. For this example I am using the letter of the region (origin) as the symbol for each measurement.

From a purely visual analysis, weight appears to be a reasonable approximation of a normal distribution, oil is bimodal, protein may be skewed left, and carbohydrate (CHO) may be either bimodal or skewed right. Tests called Goodness-of-Fit

**TABLE 11.1**
Measurements from a Sample of 12 Beans

| Bean | Weight (mg) | Oil (%) | CHO[a] (%) | Protein (%) | Origin |
|---|---|---|---|---|---|
| 1 | 102 | 10 | 42 | 43 | S |
| 2 | 94 | 13 | 49 | 33 | S |
| 3 | 96 | 12 | 45 | 39 | S |
| 4 | 91 | 14 | 52 | 29 | S |
| 5 | 97 | 22 | 33 | 40 | N |
| 6 | 96 | 24 | 34 | 37 | N |
| 7 | 100 | 22 | 33 | 40 | N |
| 8 | 98 | 21 | 32 | 42 | N |
| 9 | 97 | 23 | 30 | 42 | W |
| 10 | 97 | 21 | 29 | 45 | W |
| 11 | 93 | 24 | 26 | 45 | W |
| 12 | 100 | 20 | 28 | 47 | W |

[a]CHO = Carbohydrate.

are available to see if the data is normally distributed. The assumption of normality, however, is usually a robust assumption, which means that most tests based on a normality assumption can be used as long as the distribution is not too far off. If the distribution is far from normal, you should be able to recognize the anomaly in a scatter plot, and later when tests produce confusing results you will remember that nonnormality could be the source of the problem.

The mean and median values and estimates of modes are weight, mean = 96.75, median = 97, mode around 97; oil, mean = 19.00, median = 21, primary mode at 21–22 with a possible secondary mode around 12; protein, mean = 38.58, median = 41, mode around 41–42; CHO, mean = 37.17, median = 33, primary mode at 27 with a possible smaller mode at 45–50.

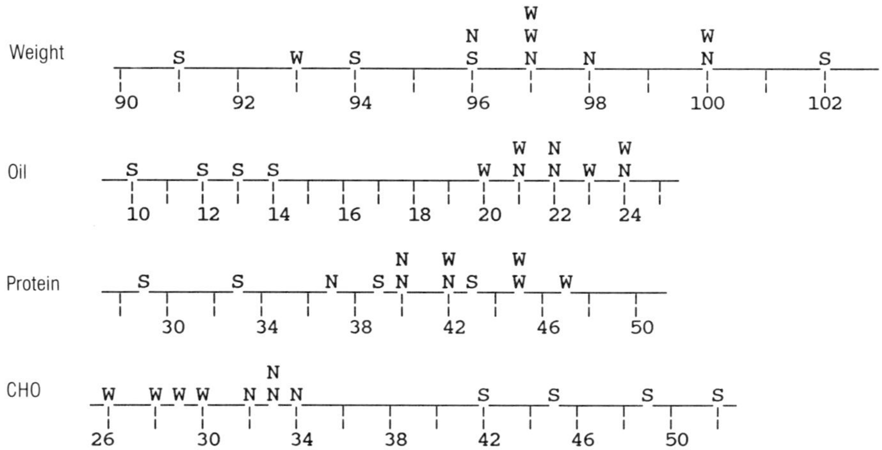


**Figure 11.5.** Scatter diagrams.

The mean, median, and mode of the weight are similar and the scatter plot looks good, so there is no reason to suspect a nonnormal distribution.

We calculate the variance by taking the difference between each observation and the mean, squaring that difference, adding up all the squares, and dividing by n–1. The sum of all the squared differences is called the sum of squares. The sum of squares divided by the degrees of freedom is the variance, and the square root of the variance is the SD. We use n–1 rather than n as a divisor because we had to estimate the mean. If the mean had been known exactly (see the probability example later), we would divide by n rather than n–1. Your calculator or computer program may ask if you want the sample variance or population variance. For research data you will want the sample variance almost every time.

To calculate the variance for weight, you will add $(102 - 96.75)^2 + (94 - 96.75)^2 + \cdots + (100 - 96.75)^2 = 106.25$ (notation: sum of squares for weight = $106.25 = \sum w^2$ ). The variance of weight is 106.25/11 = 9.659. The SD of the weight is the square root of the variance or 3.11 mg. Therefore, the population represented by these 12 beans could be expected to have 95% of the beans within a range of 90.5–103.0 mg (96.75 ± 2 SD) and 99% within a range of 87.4–106.1 mg (96.75 ± 3 SD). The standard error of the mean, the SD of the data divided by the square root of the number of observations, is 0.90. Means are usually listed with plus or minus standard errors. I am 95% sure that the "real" mean, of which 96.75 is an estimate, is between 94.95 and 98.55 [96.75 ± (1.96 × 0.90)].

Another useful measurement is the coefficient of variation (CV), 100 multiplied by the standard deviation and divided by the mean, in this case 100 × 3.11/96.75 = 3.2. The standard deviation is 3.2% of the mean. Note that CV has no units (3.11 mg/96.75 mg cancels the units), therefore CV should be the same whether we weighed these beans in kg, pounds, cwt, or whatever. Coefficient of variance is useful for comparing results across experiments, variation at different levels, and as a rule-of-thumb, early data check. If we found some beans that weighed an average of 250 mg each, then we might expect a SD of around 8 mg (3.2% × 250 mg). When the CV increases as the mean increases, then we have a "scale effect" and the larger the measurement—the more the variation. Sometimes the CV increases as the mean decreases, especially when measuring near zero and can be a sign of the difficulty of measuring accurately near zero. An experienced scientist or technician should have some rough idea of the expected CV in a measurement, and an early check of CV can be an early check for unexpected variation.

The protein mean is just a bit less than the median and the mode, indicating the possibility of the data being skewed towards lower protein contents. There are tests available for skewness and kurtosis which are based on the third and fourth moments about the mean. Often, we are not concerned with the exact amount of nonnormality but rather if it can be considered as approximately normal. The most common analytical procedure in a case where the skewness is not extreme is to proceed with the analysis as if the distribution were normal, making a mental note that if later results are confusing this skewness may be introducing some extra variation. If the extra variation is not strong enough to influence our results, this small bit of noise can be ignored.

The oil and CHO measurements are bimodal, and we must stop and consider whether we are seeing a chance random variation or if there are two separate populations. If there are really two populations, what is the underlying cause? We can see from the symbols that the beans from region S have higher CHO and lower oil. Beans from region W appear to have lower CHO. The beans from region S may have lower protein while the beans from region W may have higher protein. In subsequent sections we will discuss some probability theory and see how we can use probability to decide if differences are real or just random fluctuations.

## Probability

I like to think of probability as "applied common sense." However, people often are confused when they don't consider their question very carefully. When you ask "what are the chances of ..." or "what is the probability of ..." the question usually being asked is "of all possible things that can happen, how many of these things are ..." If I flip a coin twice and get heads twice, and then ask "what is the probability of heads on the next flip?" The answer is 50%. The coin has no memory. There are only two possible things that can happen, heads or tails, each with equal probability. If you ask "what is the probability of flipping three heads in a row?" then that is an entirely different question. All possible outcomes are: HHH, HHT, HTH, THH, HTT, THT, TTH, and TTT, where H = heads, and T = tails. Of the eight possible outcomes, only one is HHH, or three heads in a row. One out of eight is 0.125 (or 12.5%). The probability of two heads and one tail is three out of eight (HHT, HTH, and THH) or 0.375. The probability of one head and two tails is three out of eight (HTT, THT, and TTH) or 0.375 and the probability of three tails is one out of eight (TTT) or 0.125. If all probabilities don't add up to one, then you have either miscalculated or forgotten a sequence.

The "binomial formula" in any statistics book can be used to calculate these probabilities directly, but sometimes it is reassuring to go through an exercise like this and see that nothing magical or extremely complicated is occurring. If we have the time and the inclination, we can extend our probability exercise to more flips of coins, dice, gender distributions, and lottery numbers, for example. Chemists don't often spend much time flipping coins, but they do have occasion to use binomial probabilities. (Binomial means that there are two possible outcomes, i.e. heads or tails, yes or no, male or female, and above or below.)

If you are testing a method or a piece of equipment, you might use a binomial probability. Let us say that you have a known standard and you are trying to see if your new method or instrument will correctly measure that standard. You will measure how close you can get to the known value, but you should also note whether you are consistently above or below the standard. What if you try four estimates and all four are overestimates? Is your method or instrument biased? If there is no bias, then an overestimate is just as likely as an underestimate, and then the probability of an overestimate is 50% and the probability of an underestimate is 50%. The probability of getting four overestimates in a row is $0.5 \times 0.5 \times 0.5 \times 0.5$ or $1/2^4$ or 0.0625. Therefore what has just happened has a 0.0625 probability of occurring. Try this exercise for yourself. There are 16 possible combinations of over- and underestimates, of which one is all over, four are three over and one under, six are two over and two under, four are one over and three under, and one is four underestimates. A result of 0.0625 would concern but not convince me. I would try six or more estimates in the first place. If all six results are overestimates, I would be concerned that my method or instrument is biased because if it is not biased, then what I have just seen should have occurred only once in 64 times ($1/2^6 = 1/64 = 0.016$). If no bias exists, then I have just witnessed a rare event. The probability value tells me how rare.

We may use a binomial probability even when the probabilities are not 50/50. Whatever the distribution, the reasoning behind the probability is usually similar. If we have a normal distribution, we can calculate how much of the data is within or outside of a range from the mean. In a normal distribution 67% of the data is within 1 SD of the mean and 95% of the data is within 1.96 SD of the mean. The probability of drawing two observations at random from a normal distribution and finding them both to be greater than 1.96 SD from (either above or below) the mean is $0.05 \times 0.05 = 0.0025$. The probability that both would be within 1.96 SD of the mean is $0.95 \times 0.95 = 0.9025$. The probability of one within and one outside is $(0.05 \times 0.95) + (0.95 \times 0.05) = 0.095$.

When we say that 5% of the data is beyond 1.96 SD of the mean, we assume that 2.5% are 1.96 SD below and 2.5% are 1.96 SD above the mean. If we are concerned with only data above or only data below the mean, then we can use 1.645 SD. Five percent of the data are beyond 1.645 SD above the mean and 5% are 1.645 SD below the mean. When we are concerned about both above and below the mean we use a "two-tailed" statistical test and when we are concerned about only above or only below we use a "one-tailed" statistical test.

The $z$ and $t$ statistics are based on the normal distribution. Both $z$ and $t$ can be thought of as the number of standard deviations from the mean. The $z$ statistic is used in large data sets while $t$ is an approximation of $z$, which takes into account the uncertainty present when working with smaller data sets.

The $F$ and $\chi^2$ (Chi-Square) statistics are used to test variances and are not normally distributed.

### *Example 2: Probability*

We have a bag containing 60,000 beans. We have measured the oil content of every bean without damaging it. The average oil content is 19.13% and SD is 2.87%. The oil content of the beans is distributed in a perfect normal distribution (Figure 6). Thus we know that 30,000 of the beans have an oil content greater than 19.13% and 30,000 beans have an oil content less than 19.13%. We know that 40,800 beans have an oil content between 16.26 and 22.00% (16.26 = 19.13 - 2.87, 22.00 = 19.13 + 2.87). We know that 57,000 beans are between 13.50 and 24.76% (± 1.96 SD contains 95% of the beans). There are 1500 (2.5% of the total) beans with greater than 24.76% oil and 1500 beans with less than 13.50% oil. If we are concerned only with the lower 95% of the beans, then we have the 57,000 beans with less than 23.85 (all beans below the mean and all those up to 1.645 SD above the mean make up 95% of the beans. We see that 1500 beans have an oil content between 23.85 and 24.76.

We can do similar calculations for any proportion of beans. We know the mean, the standard deviation, and the normal distribution so we can easily calculate how many beans are within any range of oil content.

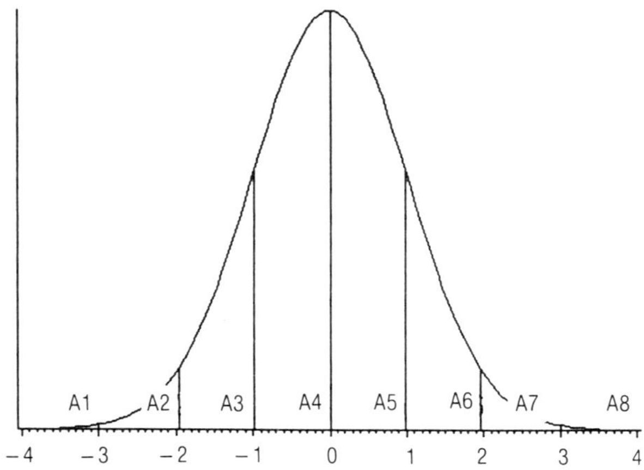


**Figure 11.6.** Bean distribution.

If the beans are distributed in a perfect normal distribution and the lines in Figure 11.6 divide at exactly −3, −2, −1, 0, 1, 2, and 3 SD from the mean, then area A1 has 78 beans, area A2 has 1,290 beans, area A3 has 8,154 beans and area A4 has 20,478 beans.

I want to find a bean with an oil content greater than 23.85%. I classify beans as "H" if they have an oil content greater than 23.85% and "N" if they have an oil content less than 23.85%. We know that 3000 (5%) of the beans are in the H category. If I pick 5 beans at random, what is the probability of getting at least one H bean? Note that I asked "at least one" and not "exactly one." Remember, these are two separate questions. The probability of getting the H bean on the first pick and then N beans on the next four picks is $0.05 \times 0.95 \times 0.95 \times 0.95 \times 0.95 = 0.04$. The probability of getting the one H bean on the first, second, third, fourth, or fifth pick is therefore $5 \times 0.05 \times 0.95^4 = 0.20$.

For you perfectionists, I will acknowledge that the probabilities are not completely stable. As I pick beans out of the bag the fractions change. The probability of picking an N bean on the first pick is 57000/60000 = 0.95. After this pick, the probability of picking a second N bean is 56999/59999 = 0.949999. I consider this difference to be trivial, and I will ignore it. However, you perfectionists can do exact calculations. If your original number is small or you are picking many beans, you should consider the changes in probability. To determine how "small" is small and how "many" are many, do a couple of the appropriate fractions on your calculator and see what the differences are. You can keep the probabilities stable by "sampling with replacement." Every time you pick a bean, replace it and remix the bag before you pick another bean.

The probability of picking five N beans in a row is $0.95^5 = 0.77$. Therefore, the probability of picking more than one H bean in five picks is $1.00 - 0.77 - 0.20 = 0.03$. You can prove this to yourself by calculating probabilities for 2, 3, 4, and 5 H beans in five picks.

## Hypothesis Testing

Hypothesis testing consists of setting up a possibility (the null hypothesis, Ho:) and calculating the probability that what has just occurred fits that possibility. If we calculate that what we have just seen is rare, then we assume that our possibility was not correct and another possibility (the alternative hypothesis, Ha:) has occurred. If we get six overestimates in a row, then we assume that the possibility that our method is unbiased is not the true situation (we reject the null hypothesis) and another possibility is likely, in this case the possibility that our method is indeed biased.

A common source of confusion is the use of the low probability to express a difference. When we say that A was greater than B ($p < 0.05$), what we are saying is that we have calculated the probability that A and B are alike or similar as being less than 5%. We set up a null hypothesis that A = B (i.e., they are two samples from the same population), and then calculated the probability of picking twice from the same population and getting a difference between A and B. A difference this large was possible but would have been a rare (< 0.05) event, and therefore we decide the null hypothesis probably is not correct. The null hypothesis is worded "is" or "equals" and the alternative hypothesis is worded "is not," "does not equal," "is greater than," and "is less than," for example.

If we start with two sets of beans from the same population and subject each of them to a different treatment, then we will test the null hypothesis that they are still from the same population vs. an alternative hypothesis that one or both of the treatments has affected the beans in such a way that they can no longer be considered similar. If they are not similar, then they cannot be considered to be from the same population. We will assign a probability value ($p < 0.05$, $p < 0.01$, or $p = 0.xx$) that they can still be considered to be in the same population. The steps to hypothesis testing are

1. Set up a null hypothesis, Ho: vs.;
2. Set up an alternative hypothesis, Ha:;
3. Decide on a statistic (i.e., $F$, $t$, or chi-square);
4. Calculate the statistic;
5. Estimate the probability of the statistic based on your null hypothesis; and
6. Accept or reject the null hypothesis.

### *Example 3: Hypothesis Tests*

Your boss gives you a bag (bag A) full of about 1,000 beans and asks you if they could have come from the bag of 60,000 (bag 0) that was used in our probability example. You pick 12 beans at random out of bag A to measure the oil and obtain 14.7, 15.2, 16.0, 16.5, 17.2, 17.3, 17.6, 18.0, 18.5, 19.1, 19.5, and 19.8% oil. You quickly calculate the mean as 17.45% and SD as 1.64%. You know that bag O had a mean of 19.13% and SD of 2.87%. This is a bit of an unusual case in research. We actually know some of the real population parameters rather than having to use estimates. We have actually measured every bean in bag O. You couldn't possibly measure every bean in a carload, in a state or region, or in a variety or biotype where we usually want to make comparisons. When you have measured every bean you have actual parameters, and you use Greek letters for them. The mean for bag O is $\mu$ and SD is $\sigma$. For bag A you have a sample of 12, which you are using to estimate the parameters. Estimates of parameters are called statistics and for bag A the mean is $\overline{X}$ and SD is s. To keep the association as clear as possible, we will use subscripts when we refer to bag A: the mean of A is $\overline{X}_a$ and SD is $\sigma_a$. We will test the hypothesis that bag A could be a sample of bag O using the six steps of hypothesis testing.

*Test 1.* First, we will compare means.

1. Set up a null hypothesis, in this case: Ho: $\overline{X}_a = \mu$
2. Set up an alternative hypothesis: Ha: $\overline{X}_a \neq \mu$ , We want to consider both $\overline{X}_a > \mu$ and $\overline{X}_a < \mu$ so we will use two-tailed tests.
3. We can use a *z*-test because we know some of the parameters. If we had to estimate all the parameters, we would have to use a t-test. See the example later in this section.
4. $z = \dfrac{\overline{X} - \mu}{\sigma/\sqrt{n}}$

   The denominator ($\sigma/\sqrt{n}$) is the standard error of the mean for samples with 12 observations. Remember, the SD is a measure of the dispersion of the data and the standard error (the standard deviation of the mean) is a measure of the dispersion of the mean. If we took many samples of 12 beans each from bag O and calculated the mean for each of our many samples, those many means would have a SD of $\sigma/\sqrt{n}$. The $z$ will be a measure of how many standard errors our $\overline{X}_a$ was from $\mu$. In our example, $z = (17.45 - 19.13)/(2.87/\sqrt{12}) = -2.03$.
5. We can check our calculated z value in two ways. In the old precomputer days, we would look up a value of z at a specific probability level (find a "critical value") and then see if our calculated value is greater than the "critical value." The value of $z$ associated with $p < 0.05$ is $\pm 1.96$. We consider $\pm 1.96$ because we are checking either direction from the mean in our two-tailed value. Our calculated value of $-2.03$ tells us that the $\overline{X}_a$ value is 2.03 SD from $\mu$ and is therefore a rare event because it will occur less than 5% of the time.

   With our new computer, we often see an exact calculation of the p value. In this case the p value is 0.042. Instead of saying "for $p < 0.05$" we can say "for $p = 0.04$" and make the same conclusions.
6. We reject Ho: thinking that 0.04 is too rare, therefore we will consider Ha: and conclude that bag A was not a random pick out of bag O.

There is nothing magic about 0.05 or 0.01. They have been used as guidelines for many years. You should consider the consequences of being wrong when you decide what level to use. If you are going to halt production and spend millions on a change in a process, then you should be more certain than 0.05.

*Test 2.* We can also compare the variances of bag A to bag O using our six steps.

1. Ho: $s_a^2 = \sigma^2$ vs.
2. Ha: $s_a^2 \neq \sigma^2$
3. We use $\chi^2$ because variances and standard deviations do not have normal distributions.
4. $\chi^2 = (n - 1)\, s_a^2/\sigma^2 = (12 - 1)(1.64)^2/(2.87)^2 = 3.59$
5. The "critical values" for $\chi^2$ are 3.82 and 21.90. The distribution of this $\chi^2$ for sample size 12 has 95% of its values between 3.82 and 21.90. A $\chi^2$ distribution is not normal and is not symmetric. Our calculated value is 3.59 which is less than 3.82.
6. We will reject Ho: that the variances are equal.

We now have two pieces of evidence that bag A is not a sample of bag O. The mean is too low, and it doesn't vary enough to be from bag O.

*Test 3.* Your boss has another bag of beans (bag B) and asks if they could be from the same source as bag A. We test 12 beans from bag B and find 17.7, 18.2, 18.7, 19.0, 19.1, 20.6, 20.9, 21.2, 22.2, 22.4, 23.0, and 24.3% oil. We immediately set up a scatter diagram and look at our data. They look different, but let's check them. We calculate $\overline{X}_b = 20.61$ and $s_b = 2.10$. First, we will compare variances.

1. Ho: $s_a^2 = s_b^2$ vs.
2. Ha: $s_a^2 \neq s_b^2$
3. We use $F$ to test the difference of two estimates of variance. $F = s_b^2/s_a^2$. An $F$ ratio always compares variances rather than SD and always tests the larger divided by the smaller.
4. $F = 4.39/2.69 = 1.63$ with 11,11 degrees of freedom.
5. The critical value for $F$ at $p < 0.05$ and 11,11 degrees of freedom is 2.82.
6. Our calculated $F$ is less than the critical value, so we will not reject Ho, and we will assume that $s_a$ and $s_b$ are similar if not equal.

We can now compare the means for bags A and B.

1. Ho: $\overline{X}_a = \overline{X}_b$ vs.
2. Ha: $\overline{X}_a \neq \overline{X}_b$
3. We use a *t*-test to compare two means. The *t*-test is called Student's *t* after Gossett who published under the name of Student. He was a chemist in an English brewery who was interested in statistical tests both as a part of his job and also for academic interest. He realized that a *z*-test did not allow for uncertainty when estimating with smaller numbers of samples. The *t*-test is a *z*-test adjusted for small samples. The *t*- and *z*-tests are alike at $n = 30$ or more.

4. The $t$-test is

$$t = \frac{X_a - X_b}{s_p \sqrt{1/n_a + 1/n_b}}$$

where $n_a$ and $n_b$ are the numbers of observations in A and B and

$$s_p = \frac{s_a^2(n_a - 1) + s_b^2(n_b - 1)}{n_a + n_b - 2}$$

$s_p$ is a pooled estimate of $s$. We don't know $\sigma$, so we estimate it with $s$. We have two estimates of $s$, $s_a$ and $s_b$. We have already tested and found them similar, so we will pool them and call the result $s_p$.

In our example:

$$s_p^2 = \frac{(1.64)^2(12 - 1) + (2.10)^2(12 - 1)}{12 + 12 - 2} = 3.5498$$

$$s_p = \sqrt{3.5498} = 1.884, \text{ and}$$

$$t = \frac{17.45 - 20.61}{1.884(1/12 + 1/12)^{1/2}} = -4.11$$

5. The critical value for $t$ with 22 degrees of freedom is 1.72 and we ignore the sign. Our calculated value of 4.11 is greater than 1.72. Our computer gives an exact $p$ value of 0.0005.

We reject Ho:. Although the beans from bags A and B have similar variances, their means are far enough apart that we will assume that they are from different origins or have been treated differently.

## Correlation and Regression

Two continuous variables are correlated if they are related in a linear manner. The most common correlation, $r$, is Pearson's or the product-moment correlation and is a unitless value that can range from −1 to 1. An $r$ of −1 between $X$ and $Y$ means that every time $X$ increases, $Y$ decreases and the reverse. An $r$ value of 1 means that every time $X$ increases, $Y$ also increases, and every time $X$ decreases, $Y$ also decreases. An $r$ value of 0 means that fluctuations of $X$ have absolutely no linear relationship with any change in $Y$. In my opinion, the correlation coefficient is the most misused and misunderstood statistic in scientific research. First, a correlation measures only linear relationships. Two variables can have a strong nonlinear relationship, such as sigmoidal, parabolic, and logarithmic, and not have a high correlation. Second, the correlation coefficient can be influenced by a very few oddball measurements at the upper or lower ends of the distribution. Third, a statistically significant correlation is often interpreted as a biologically or chemically significant correlation which is not always true. Statistical significance of a correlation refers only to the hypothesis that the correlation coefficient is zero and in no way is an indication whether the correlation is weak, moderate or strong. Chemically trivial correlations from a large data set are often statistically significant, and potentially important correlations from small but highly variable data sets are seldom significant. Finally, correlations have no cause-and-effect connotations, are useless for predictions, and can be spurious if *X* and *Y* are both related to a third variable *Z* but not directly to each other.

Correlations are, however, useful as a tool in initial data analysis. An experienced scientist knows which variables should be correlated and which should not and knows that further investigation is necessary when correlations do not follow expectations.

The first and most important rule when comparing two variables is "Look at your data!" With modern computers and graphical software there is no excuse for accepting a correlation coefficient and not looking at the data. A quick visual appraisal can often discover nonlinear relationships, prominent outliers, and the relative strength of the relationship.

A regression coefficient also is a measure of a linear relationship, but is much more useful than a correlation coefficient. Unfortunately, there is no standard notation for simple linear regression equations. Common notations are: $Y = mX + b$, $Y = a + bX$, and $Y = b_0 + b_1X$. Each has the meaning that a response (or dependent) variable (*Y*) can be predicted from an intercept value ($b$, $a$, or $b_0$, which is the value of *Y* when $X = 0$) plus a slope or regression coefficient ($m$, $b$, or $b_1$) multiplied by an independent variable (*X*). (I prefer the notation, $Y = b_0 + b_1X$, because the step to multiple regression, $Y = b_0 + b_1X_1 + b_2X_2 + \cdots + b_nX_n$, is simpler.) A regression line is an excellent point of reference for judging bias, nonlinearity, and the influence of extreme values when you are examine your data.

The regression approach does have some caveats: There is an implication that a change in *X* is responsible for a change in *Y* (i.e., a cause and effect). The regression coefficient is the amount of change in *Y* due to a one unit positive change in *X*. You also assume that *X* is measured without error and all of the uncertainty is in *Y*. This assumption can cause problems when a method or instrument is calibrated over a range of known *X* values and then used to estimate *X* from a measured *Y*. The correct procedure is complicated, and you should consult a statistician or statistical manual before you try to estimate SD of *X* predicted from *Y*.

### *Example 4: Correlation and Regression*

We will use the data from Table 1. First, we will calculate the correlation between weight and protein. We need the sums of squares for both weight and protein, which we calculated when we estimated the variance for each ($\sum w^2 = 106.25$ and $\sum p^2 = 295.67$). We calculate the sum of cross-products (notation $\sum wp$) by multiplying the difference between the weight measurement and its mean by the difference between the corresponding protein measurement and its mean. If both weight and protein in a bean are larger than the mean for all beans or both are smaller, the cross-product will be positive. If one is larger and the other smaller, the cross-product will be negative. The sum of cross-products divided by the degrees of freedom $(n - 1)$ is the covariance.

For weight and protein we calculate $(102 - 96.75) \times (43 - 40.167) + (94 - 96.75) \times (33 - 40.167) + \cdots + (100 - 96.75) \times (47 - 40.167) = 109.500 = \sum wp$. The formula for the correlation coefficient, *r*, is

$$r = \frac{\sum wp}{\sqrt{(\sum w^2 \sum p^2)}}$$

Note that the units (mg × %) in the numerator cancel with the units (mg × %) in the denominator, thus $r$ is unitless.

In this example, $r$ for weight and protein is

$$r = \frac{109.50}{\sqrt{106.25 \times 295.67}} = 0.618$$

You can look in a table and find that for 11 degrees of freedom, the $r$ value needs to be 0.55 or greater to be considered significant at $p < 0.05$. You can calculate a probability directly by estimating t and then looking up the $p$ value for that $t$. The correlations for the other measurements in Table 11.1 are

$$\begin{aligned} r_{\text{weight,CHO}} &= -0.37\ (p = 0.24) \\ r_{\text{weight,oil}} &= -0.01\ (p = 0.98) \\ r_{\text{CHO,protein}} &= -0.83\ (p < 0.01) \\ r_{\text{CHO,oil}} &= -0.84\ (p < 0.01) \\ r_{\text{protein,oil}} &= \ \ 0.40\ (p = 0.20) \end{aligned}$$

The regression of protein on weight is calculated by

$$b_1 = \frac{\sum w\text{p}}{\sum w^2} = \frac{109.50}{106.25} = 1.03$$

The regression coefficient has units, in this case %/mg. As weight increased 1 mg, the protein content increased 1.03%.

We have estimated that Protein = $b_0 + b_1$(Weight). If we use the mean values for Protein and Weight, we can calculate $b_0 = 40.167 - 1.03(96.75) = -59.49$ which theoretically is the value of $Y$ (Protein) when $X$ (Weight) = 0. In this case such a value is unlikely, but we still need the intercept value in order to properly place the regression line on our graph.

Statistical tests are available to see if either $b_0$ or $b_1$ are different from zero. Most computer stat packages list them. If the correlation is significant, you can assume that the slope will probably be significant. The correlation coefficient squared, usually listed as $R^2$, is called the "Coefficient of Determination" and can be interpreted as the amount of variation in $Y$ that can be explained by differences in $X$. In this example, we can say that 37% ( $= 0.62^2$) of the differences in protein in these beans were due to differences in weight.

Other regression equations to predict oil % in these beans are

Oil % = 20.20 – 0.0141 (Weight), $R^2 = 0.00$
Oil % = 3.03 + 0.3935 (Protein), $R^2 = 0.16$
Oil % = 36.54 - 0.4908 (CHO), $R^2 = 0.70$

Multiple regressions are possible, but are beyond the scope of this chapter. If you do want to use multiple regression, for best results the predictor variables should be independent of each other. In the above example, if I wanted to predict Oil % with Weight, Protein, and CHO considered simultaneously, I should be concerned that Weight, Protein, and CHO are all related to each other as well as to oil. The closer the relationships, the more unstable the regression coefficients are. A change in one predictor variable affects not only the response variable (oil) but also the other predictors.

## Analysis of Variance

Analysis of variance (ANOVA) is a common method of partitioning variance (separating signal from noise). Mathematical procedures are used to measure the variance in a response variable ($Y$) and then assign portions of it to the possible sources. Sources can be as simple as one variable with two levels (Example 5) and can be as complicated as computer facilities and interpretative knowledge will allow. The ANOVA is a very powerful method of examining the effect of one or more treatments in the presence of other influences plus nuisance variables. Sources can be qualitative or quantitative, fixed or random, independent or related, or even wanted and unwanted.

A quantitative source of variation, such as weight, time, and concentration, can be included as a continuous variable in an analysis of covariance. The procedure will estimate the influence of a continuous variable on the response variable while mathematically holding the other sources of variation constant.

A source which represents all possible levels or examples of an entity is a fixed source. A random source is one which contains a random sample of all possible levels or examples. Year is often a random source when you wish to use results from several years to make predictions for other years. If you are testing individual kernels of corn from several different varieties and you are using two or more kernels from a individual ear, then the ear is a random source. The question of random vs. fixed is seldom simple but is important because the unacknowledged influence of random sources can cause misleading results. Random and fixed sources can be analyzed together in the dreaded Mixed Model Analysis of Variance.

Two or more sources are said to interact if their effects are synergistic. An elementary example is the effects of three different meals on the growth rates of mice. The control mice ate a diet with a minimum amount of protein and energy. The other mice ate the control diet plus supplemented meal. Meal A contained little protein but many calories. Meals B and C contained few calories but a great deal of high-quality protein. Mice that ate either A, B, or C alone gained weight a bit faster than the controls. Mice that ate A and B, A and C, or A, B, and C gained weight very rapidly. Mice that ate B and C together did not do any better than those eating B or C individually. The meals, therefore, interacted. We must be careful in interpreting simple effects in the presence of interactions. If asked "does meal C result in an increase in growth?" we would have to answer "a bit by itself or in combination with meal B, but a great deal when combined with A." Interactions are common in nature and are an important reason why experiments should be designed to detect and measure interactions. If many factors are to be tested or screened, a design with all possible interactions and replications can be large. A statistician can help save time and resources with fractional factorial designs which take advantage of partial replications.

### *Example 5: Analysis of Variance*

The fundamental concept of Analysis of Variance is that the total variance of a measurement can be divided into known and unknown causes. The known causes are called treatments or sources of variation, and the unknown causes are called error or residual variance. The term residual is used because the error is the residual portion after all other possible sources of variation have been estimated.

We will use the % oil data from Table 11.1. We will test to see if location made a difference in oil content. We know from our calculations for estimating the variance that the total sum of squares (SST) for oil % is 283.667. Next we calculate the mean % oil for each location, which are $S = 12.25$, $N = 22.25$, and $W = 22.00$. Within each location we calculate a sum of squares based on the mean for that location. For $S$ we calculate $(10 - 12.25)^2 + (13 - 12.25)^2 + (12 - 12.25)^2 + (14 - 12.25)^2 = 8.75$. For $N$ we get 4.75 and for W we get 23.50. Thus, the sums of squares within the locations add up to 23.50, which is called the sum of squares for error (SSE). Next we calculate the sum of squares between locations, which is the squared differences between location means and the overall mean. $(12.25 - 18.83)^2 + (22.25 - 18.83)^2 + (22.00 - 18.83)^2 = 65.04$ which we have to multiply by the number of observations in each location $= 65.04 \times 4 = 260.17$, which is the treatment sum of squares (SSA). If we have done our calculations correctly, SSA and SSE will add up to SST, $260.17 + 23.50 = 283.67$. You should be able to recognize all of the information that the computer gives you.

We divide the SSA by the degrees of freedom for treatments $(t - 1)$ to find the Mean Square for Treatments (MSA), MSA = 260.17/2 = 130.08. The SSE is divided by the number of treatments multiplied by one less than the number of observations in each treatment $[t(n - 1)]$ to find the Mean Square Error (MSE), MSE = 23.50/9 = 2.61. The MSA contains the error variance, $\sigma_e^2$, plus $n$ multiplied by the variance due to treatments, $\phi_t$. The MSE contains $\sigma_e^2$. Thus, if there is no variance due to treatment, the $\phi_t$ term will be near zero and the ratio MSA/MSE will be one. The probability distribution associated with this ratio is an $F$-distribution with $(t - 1)$ and $k(t - 1)$ degrees of freedom. In this case MSA/MSE = 130.08/2.61 = 49.82. You can look up $F_{2,9}$ in a table in any basic statistics book and find a critical value of 19.4 at $p < 0.05$. If $\phi_t$ is really zero then an $F$ of 49.8 is quite rare (< 0.05), and you will probably reject the hypothesis that the treatment effect is really zero.

The analysis of variance tells you only whether a difference between treatments exists. It does not tell you which treatments are different from each other if there are more than two treatments. In order to tell more about the individual differences, you will need to examine some individual comparisons. Multiple comparisons, or multiple range tests, are a subject of great controversy among statisticians. The least significant difference (LSD), Duncan's, Sheffé's, Tukey's, Waller's, Bonferonni's, and Student-Newman-Keuls' tests, as well as pairwise $t$-tests of least squares means are all used and misused. With three treatments, as in our example, the results should be similar no matter which test you use. The differences among the various tests grow larger as the number of treatments increases. Remember that you are looking for rare events and rejecting the possibility of those events because of their rarity. If you are comparing each of eight treatments to each other, you are making a total of 28 comparisons between treatment means and to find one significant difference would not really be a rare event. The LSD test is the simplest and makes no allowances for the number of treatments. The pairwise *t*-test of least squares means is useful for making specific planned comparisons. The objective of most of the other tests is to make adjustments for the number of means in order to keep rare events really rare. Often, I have found that the arguments about which test to use are unnecessary if proper planning is made. Many times a group of 8 or 10 means can be divided (blocked) into 2 or 3 sets of means based on origin; region; or even treatment similarities, such as chemical family, base, or concentration. If you can block on a higher level and then test four or fewer means within each block, you not only avoid multiple range ambiguities but you can make more general statements about your results.

Most of you will use a calculator or computer which will do all of the calculations. Run this example and make sure that you recognize the results in your program or software. There are calculation formulas which make these estimations less tedious and I encourage you to work through a simple analysis of variance just to see what is happening.

## Multivariate Analysis

Multivariate analysis is based on the assumption that measurements of any one variable are redundant to some extent because the same information affects several variables. The objective of multivariate analysis is to reduce the total number of variables needed to describe information on the objects in question.

Procedures such as Principle Components Analysis, Factor Analysis, Canonical Correlations, Profile Analysis, Partial Least Squares, Discriminant Analysis, and some others are examples of multivariate analysis. All attempt to reduce very complicated data relationships to simpler and more understandable levels. I cannot begin to explain these complicated procedures in detail in this chapter (indeed, I could not explain some satisfactorily at all). I will give a brief explanation of some of the more common procedures.

The problems with multivariate analysis stem from the difficulties of the mathematics. Matrix algebra tends to be difficult to understand conceptually, and the calculations are tedious. The assumptions associated with multivariate procedures along with the consequences of violating those assumptions are not always well understood. The results are often in vectors rather than single numbers and can be difficult to interpret. As computers alleviate the tedium of calculations and allow much more experimentation on multivariate theory, I expect multivariate procedures to become more common and as applied scientists use these procedures, I expect interpretations to become more common and more understandable.

Multivariate analysis involves several dimensions or measurements on the same individuals. For example, we can measure time, pressure, temperature, and solvent concentration during the treatment of a bean and later measure several different chemicals within that bean. These measurements tend to be related to each other, and we realize that we really can't study each measurement independently of all the others.

A relatively simple multivariate technique is partial correlation. In a previous example, we calculated correlations between CHO and protein at –0.83, CHO and weight at –0.76, and protein and weight at 0.84. We can see that these three measurements are interrelated. We might wonder if CHO and protein are related independently of weight or if the correlation of –0.83 is simply a reflection that both are related to weight. The partial correlation coefficient between CHO and protein while holding weight constant is –0.54 (Notation: $r_{\text{CHO,protein.weight}} = -0.54$). We conclude that CHO and protein are related to each other independent of their common relationship with weight. The calculation of partial correlation is relatively simple. To estimate the partial correlation between $X$ and $Y$, holding $Z$ constant, find $r_{XY}$, $r_{XZ}$, and $r_{YZ}$ and

$$r_{XY.Z} = \frac{r_{XY} - r_{XZ}r_{YZ}}{\sqrt{(1 - r^2_{XZ})(1 - r^2_{YZ})}} \quad .$$

Analysis of variance in a multivariate form is called MANOVA. When multiple interrelated sources of variation affect multiple responses, MANOVA is an option. The MANOVA is not always an advantage over univariate ANOVA unless the variables are highly related. The repeat measures problem discussed in the next section is often approached by multivariate procedures because of the high degree of relationship between adjacent measurements.

A Profile Analysis is a useful procedure that should be more common. Several responses are often measures over a gradient, such as time or concentration. These responses are in the form of two or more lines of any shape and are difficult to interpret or compare. The Profile Analysis will test in three steps. First, are the lines parallel? (They don't have to be straight to be parallel.) Second, if they are parallel, what is the probability that they are variations of the same line? (i.e., are they at the same levels throughout the gradient?) If steps one and two find that you are looking at variations of one line, what is the probability that the one line is flat? Good software for a simple profile analysis is difficult to find, but the process is well described by Morrison in *Multivariate Statistical Methods.*

Discriminant Functions are a widely used in classification problems. Microbiologists might try to derive a function of several measurements in order to place newly discovered microbes in different species. College administrators have used discriminant functions to use all available information in deciding whether or not to admit students.

Canonical Correlations are used to estimate dependencies between two sets of data. As an example, suppose we have processed oils at different temperatures and pressures and stored them for various lengths of time. We then measured flavor scores, odor, and color for each sample. Our first canonical correlation is 0.65, which we use to derive the canonical responses:

$U_1$ = 0.2(temperature) + 0.1(pressure) + 1.0(storage time), and
$V_1$ = –0.8(flavor) + 0.2(color) + 1.0(odor).

Our second canonical correlation is 0.20, which we use to derive the responses:

$U_2$ = 0.5(temperature) − 1.0(pressure) + 0.1(storage time), and

$V_2$ = −0.2(flavor) + 1.0(color) + 0.1(odor).

We see that the first and most important relationship is between storage time and a combination of flavor and odor. The second relationship is the effect of pressure and temperature, with temperature having one-half of the effect pressure has on color. The interpretation of the results of canonical correlations relies on your ability to relate vectors, which can be quite complicated.

Factor Analysis, of which Principal Components is a subgroup, is an attempt to find hidden factors, latent variables, which have generated the dependencies or variations in your measured responses. Gould explains the development, use, and misuse of the theory and applications of factor analysis by psychologists searching for a measure of general intelligence in *The Mismeasure of Man*. This book is highly recommended to anyone interested in multivariate analysis, intelligence testing, or the history of science in general. Gould originally became interested in factor analysis in his attempts to define the general shape and size factors of fossils. A chemist might use factor analysis to examine the combined effects of a series of volatile compounds as they affect flavor or odor. Once combinations of the volatiles are defined, the chemist can look for common subunits, chains, or molecule families.

Partial Least Squares is a newer technique which attempts to select combinations of variables simultaneously in both the predictor and the response data sets.

## Planning Experiments

The two greatest problems in modern research are the lack of a clear objective, and an unknown or incorrect inference space.

Objective statements in most of the manuscripts or research proposals that I read tend to be too general. A specific well-written objective statement not only tells the reader what to expect but, perhaps more importantly, serves as a point of reference for the researcher. Experiments have a tendency to drift in different directions over time and especially as preliminary results are evaluated. A statement such as "we propose to study the correlation between location and quality" may sound good at first, but it will become a source of frustration before the experiment is completed. Does location refer to state, geographic area, land slope, or some other criteria? What makes these locations different? How can I measure these differences? Quality is in the eye of the beholder. Do you mean quality to a grower, processor, or end-user? Be specific about what is actually going to be measured.

I consider the analysis, interpretation, and presentation of results as integral parts of any experiment. The laboratory work is only the first half of the experiment. As in other activities, the job is not over until the paperwork is done. Too often data are collected with no idea of how they will be analyzed. I recommend that you draw up graphs and tables of your expected results before you ever start the experiment. I have found that you can work backwards to find the most practical and efficient procedures. If you have a reasonable and understandable presentation of your results in mind, then the analysis was probably reasonable and understandable, and the data were collected in a reasonable manner. Data that are collected with no clear idea of how they will be presented are often difficult to analyze and will lead to large, unwieldy tables and incomprehensible graphs. Multivariate analyses are often a last resort because no simpler analysis was planned. Neither the scientist nor the consulting statistician likes to conduct "salvage operations" on large data sets which appear to have been collected as a demonstration of a new method or instrument.

Think through your entire experiment before you ever start it. What question is the experiment asking? Be specific! What will you know at the end of the experiment that you don't know now?

Think through your variables. Can you measure what you propose to measure? How are you going to assign a number to "quality," "bigger," "better," or "longer"? How many treatments do you need and at what levels? How fine or large a difference (resolution) is important, and can you measure at that level? Can you calibrate instruments and people at the levels where you want to collect data? Will your treatments interact with each other? Are nuisance variables that you can't control, such as time, temperature and humidity, sample source, or instrument differences, a potential problem? You can often measure these variables and correct for them mathematically if you have the foresight to measure them properly. I have been accused of suggesting "dry labbing" but I firmly believe that if you have planned a simple method of presenting your results, then the analysis will be simple and the experiment will go simply.

The "inference space" is where the results of your experiment will apply. Inference space and population are often synonymous. If you study oil content in soybeans, your conclusions will be valid for only those soybeans that are represented by your experimental sample. If you make the statement "the composition of soybean oil is ... " then you are making a broad statement about soybeans in general, and are you sure that your results would be valid in all soybeans? The reason that you are doing research in the first place is so you can make observations and predictions about your inference space; therefore you should be extremely careful about what population has been sampled. If you are gathering information about a carload, then you should carefully sample that carload. Every bean in that car should have the same chance of ending up in your sample. If you take a field sample of grain and carefully clean it before running any experiments, your experimental results apply to cleaned grain and not necessarily to field grain.

You should know the difference between a random sample and a representative sample. A random sample means that every member of the population had an equal chance of being included. If you go out of your way to make sure that you have some northern, some southern, and some from in-between, then you do not have a random sample. When you chose a southern sample, the northern and in-between samples did not have an equal chance and vice versa. You can choose at random within north, south and in-between, but you have now introduced a variable "location." This example is usually readily understood, but complications arise when small, medium, and large; or light, medium, and dark; or new, medium, and old are chosen. I am not advising you to avoid choosing representative samples. What I am advising is that you recognize when you have restricted your randomization and include the new variable, blocking, in your analysis.

An experimental unit is the unit to which the treatment was applied and the observational unit is the unit that was measured. They can be the same. However, if you roast 1,000 beans at a specific temperature and then measure several of the beans individually, the batch of 1,000 beans is the experimental unit and the individual beans are the observational units. The importance of this difference is that in order to test a treatment, the experimental unit must be replicated. No matter how many beans of the 1,000 you test, you have tested only one batch. Your inference space is that specific batch of beans and not any broader population of beans. If you repeat the process with a second batch of 1,000 beans from a second variety or source, then you can make inferences about this second batch and even compare batches. You cannot, however, make any inferences about the two varieties. Even if you test 100 beans from each batch, you have tested each variety only once. I will stubbornly insist that the differences between the two batches were due to a quirk of the oven or even because of the difference in the time of day that they were run, and you cannot prove differently unless you replicate the batches. This difference between experimental unit and observational unit is often misunderstood and can cause misleading results. A common practice is to extract a batch of beans or kernels and then take grab samples of the oil and call them replicates. I will ask you how many times could the effects of one strange bean have shown up in your results? You are not replicating anything but your technique if that one bean can show up more than once. Valid replication is often time and resource consuming but is absolutely essential for results to be applicable beyond your lab.

A common problem is to do a "repeated measures" experiment without realizing it. What if you were asked to review an experiment where one diet is fed to one person and another diet fed to a second person, and then multiple blood samples are drawn from each person over a period of time? You would immediately recognize that one person per diet is not adequate. There would be only one experimental unit per diet. You would, of course, recommend more people per diet. The second problem that is not always apparent is that the blood samples are not independent over time. If one person had something odd in their blood, that oddity would show up in all the samples from that person. Samples from the same person are related to each other more than samples between persons would be expected to be related. Statistical techniques developed under the general title of "Repeat Measures Analysis" are available and should be used. The problem is recognizing when measurements are repeat measures. If a batch of beans or a quantity of oil is heated or treated and then samples are taken over time, those samples are repeat measures. If one oven is used to heat several samples and the samples are removed at time intervals, those samples are repeat measures. If beans are bulked and ground and then grab samples are taken from the bulk lot, the lot is the experimental unit and you are taking repeat measures of it.

## Conclusions

I am occasionally discouraged when I see an experiment where the best people used the best equipment with great care and then tacked a half-baked statistical analysis on the end of their work. Careful work requires careful analysis. Plan your analysis before you start your experiment. Find a statistician or a statistical reference book that you understand before you start your experiment. Misunderstandings are the result of poor communication. Be clear about what you want to describe or compare. Understand the variance structures. Your measurements are not all the same. Why not? You can think about why measurements vary or how two measurements are related and then use statistics to explain your thoughts. Instead of using the word "significantly," use "consistently." If your results don't fit your expectations, don't accept defeat. Find out why. Were the wrong tests used? Were assumptions violated? Was there too much variation to find any consistencies (i.e., did the noise overpower the signal, or was the signal too weak to discern?) Many times I hear "If I had more samples, the results would be significant." Maybe if you had more numbers you would just have more noise, but you might be right. You should have considered the amount of difference you wanted to detect in light of the expected variation. The last thing to consider is that maybe your expectations were wrong. Fisher talked about the disappointment a scientist feels when a beautiful, well-conceived theory is confronted with an ugly fact.

Statistics are not magic. There are reasonable explanations for almost everything statisticians do. The modern computer age is beginning to change the way we use statistics and you should be prepared to consider new calculation-intensive methods. Just try to understand why you are doing what you are doing.

## References

1. Box, G.E.P., Hunter, W.G., and Hunter, J.S. *Statistics for Experimenters,* John Wiley & Sons, New York, 1978.
2. Gould, S.J. *The Mismeasure of Man,* W.W. Norton, New York, 1981.
3. Hicks, C.R. *Fundamental Concepts in the Design of Experiments,* Holt, Rinehart and Wilson, New York, 1982.
4. Massart, D.L., VanDeginste, B.G.M., Deming, S.N., Michotte, Y., and Kaufman, L. *Chemometrics: A Textbook,* Elsevier, New York, 1988.
5. Morrison, D.F. *Multivariate Statistical Methods,* McGraw-Hill, New York, 1976.
6. Ostle, B. *Statistics in Research,* Iowa State University Press, Ames, Iowa, 1963.
7. Snedecor, G.W., and Cochran, W.G. *Statistical Methods,* Iowa State University Press, Ames, Iowa, 1973.
8. Steel, R.G.D., and Torrie, J.H. *Principles and Procedures of Statistics,* McGraw-Hill, New York, 1960.
9. Walpole, R.E., and Myers, R.H. *Probability and Statistics for Engineers and Scientists,* MacMillan, New York, 1989.
10. Wernimont, G.T. *Use of Statistics to Develop and Evaluate Analytical Methods,* Association of Official Analytical Chemists, Arlington, Virginia, 1985.

# Index